From the Nation's #1 Educational Publisher K-12

The McGraw-Hill Junior Academic Series

Kindergarten

Enrichment Skills

Challenging and Fun Activities

Math • Reading

Table of Contents

Credits:
McGraw-Hill Learning Materials Editorial/Production Team
Vincent F. Douglas, B.S. and M. Ed.
Tracy R. Paulus
Jennifer Blashkiw Pawley

Design Studio
Cover: Beachcomber Studio
Interior: Color Associates

Warner Bros. Worldwide Publishing Editorial/Production Team
Michael Harkavy
Paula Allen
Victoria Selover
Charles Carney
Allen Helbig
Tim Armstrong
Holly Schroeder

Illustrators
Cover & Interior: Animated Arts!™

McGraw-Hill
Consumer Products
A Division of The McGraw-Hill Companies

Send all inquiries to:
McGraw-Hill Consumer Products
250 Old Wilson Bridge Road
Worthington, OH 43085

1-57768-500-8

1 2 3 4 5 6 7 8 9 10 WEB 04 03 02 01 00 99

From the Nation's #1 Educational Publisher K-12

The McGraw·Hill Junior Academic Series

Kindergarten

Enrichment Math

Challenging and Fun Activities

Numbers & Counting • Fractions

Time & Money • Problem Solving

Table of Contents

Table of Contents (continued)

NAME ______________________________

Counting and Matching

5

3

4

6

Directions: Have your child look at the number at the end of each row that shows how many in all. Ask your child to draw more shapes to match the number. ***Skill:*** Counting and matching.

NAME ______________________________

ADDITION

1. 

3 | 1 | 4

2. 

3. 

4. 

Directions: In the first and second columns, have your child write the number to tell how many animals are shown. In the third column, ask your child to write the number of animals in all. ***Skill:*** Practicing beginning addition.

NAME ________________________

ADDITION

3 + 2 = 5

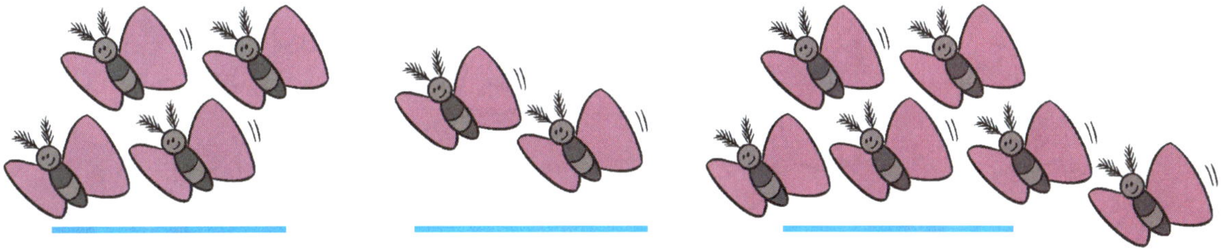

____ + ____ = ____

____ + ____ = ____

____ + ____ = ____

Directions: Have your child count the number of objects in each group, then write the number to show how many. Ask your child to write the number to tell how many in all. ***Skill:*** Practicing addition.

NAME

REVIEW

 2 + 2 = 4

 ___ + ___ = ___

 ___ + ___ = ___

 ___ + ___ = ___

Directions: Have your child find the groups of animals in the picture. Following the example, ask your child to write an addition sentence for each group. ***Skill:*** Writing addition sentences.

NAME ______________________________

ONE LESS

Directions: Have your child look at the objects in each block, then draw another group that shows one less.
Skill: Identifying one less.

NAME

SUBTRACTION

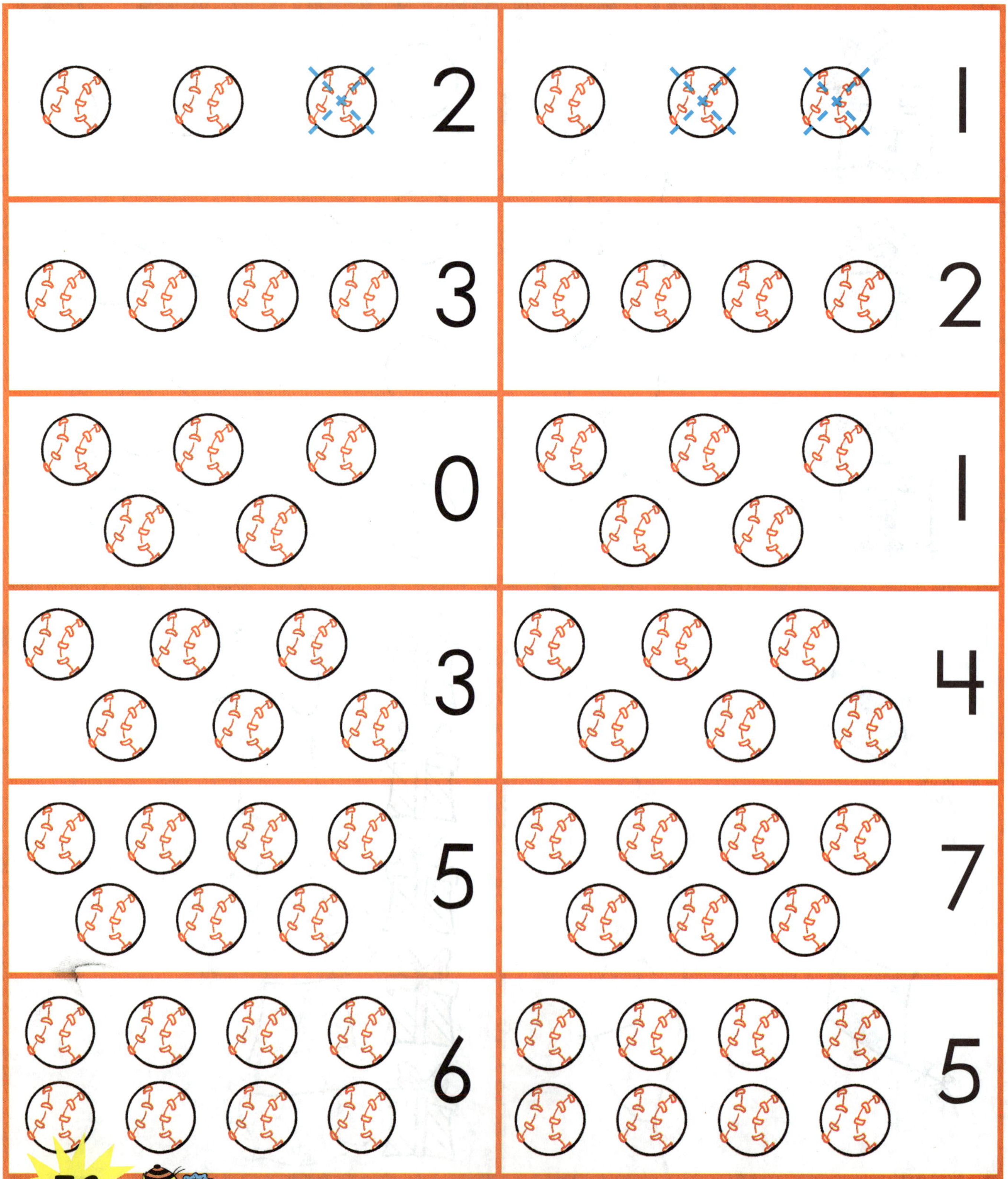

Directions: Have your child look at the number in each row, then cross out the baseballs to show how many are left. ***Skill:*** Practicing beginning subtraction.

NAME ______________________________

SUBTRACTION

1.

4 3 1

2.

3.

4.

Directions: In the first and second columns, have your child write the number to tell how many animals are shown. In the third column, ask your child to write the number of animals that are left.
Skill: Practicing beginning subtraction.

NAME ______________________

REVIEW

$5 - 2 = 3$

___ − ___ = ___

___ − ___ = ___

___ − ___ = ___

Directions: Have your child count the number of space objects in each group, then write the number to show how many in all. Then ask your child to write the number to tell how many are left.
Skill: Practicing subtraction.

NAME ____________________

PROGRESS CHECK

Directions: Have your child find the pattern in each row. Ask your child to draw what comes next.
Skill: Identifying one more and one less; identifying and completing a pattern.

NAME ______________________

PROBLEM SOLVING

1 + 2 = ____

3 + 1 = ____

2 + 3 = ____

Directions: Have your child follow the paths to the correct addition sentences. Ask your child to solve the addition sentences by writing the number to tell how many of each object in all.
Skill: Identifying and solving addition sentences.

NAME ____________________

ADDITION AND SUBTRACTION

+ →

1	1	2
1	2	3
2	3	5

– →

5	2	
3	1	

+ →

2	1	
1	1	

Directions: Have your child complete the squares by adding or subtracting the numbers in each block. ***Skill:*** Practicing addition and subtraction.

NAME ______________________________

Addition and Subtraction

3 [–] 1 = [2]

2 [ ] 2 = [ ]

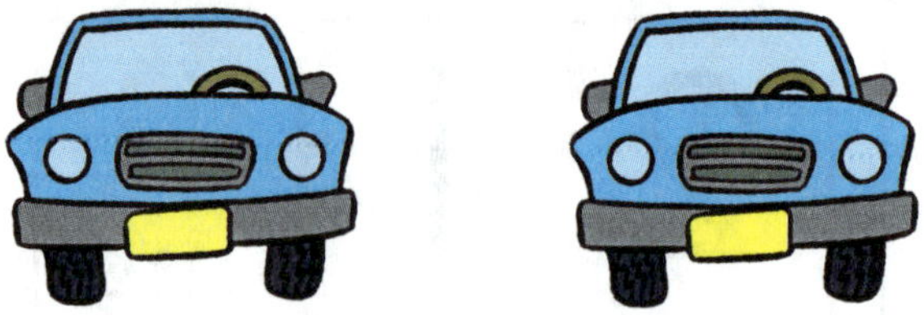

3 [ ] 2 = [ ]

–	+	+	4	2	5

Directions: Have your child look at the pictures, then read each problem. Ask your child to look at the numbers and signs at the bottom of the page, then use this information to fill in the empty squares.
Skill: Practicing addition and subtraction.

NAME ______________________________

COUNTING

4

Directions: Have your child circle the robots with square heads and place an X on the robots with round heads. Ask your child to count how many robots in all, then write the answer in the space provided. ***Skill:*** Counting and matching.

NAME

MORE THAN 10

Directions: Have your child look at the picture, then circle the objects at the bottom of the page that show more than 10. ***Skill:*** Identifying groups of 10 and more.

NAME ____________________

Numbers 13 to 16

○	13	14	15	16	⬡	13	14	15	16
△	13	14	15	16	□	13	14	15	16

Directions: Have your child look at the figures in each row, then circle the figure that is different. Ask your child to look at the shapes at the bottom of the page, count the shapes, and then circle the number that tells how many of each shape in all. ***Skill:*** Identifying groups of 13 to 16.

NAME ____________________

NUMBERS 14 TO 20

15	17	20	19	14	18

20

15

Directions: Have your child study the chart at the top of the page. Ask your child to write how many of each bird, then circle the bird in each block that has the greater number. ***Skill:*** Identifying groups of 14 to 20; identifying more.

NAME

NUMBERS 20 TO 25

22

Directions: Have your child count the animals in each group, then write the number that tells how many.
Skill: Identifying groups of 20 to 25.

NAME

NUMBERS 26 TO 30

28

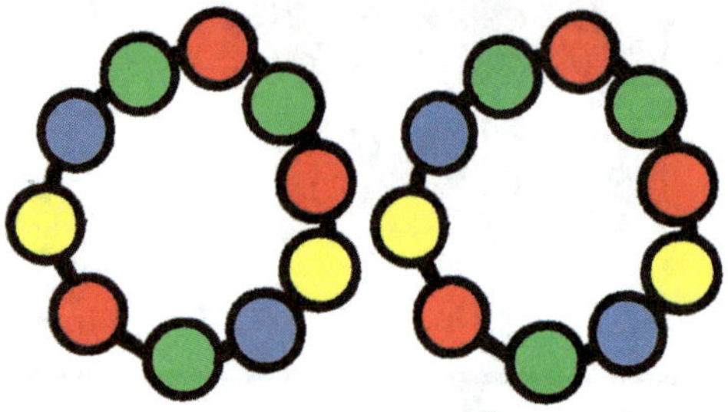

28

29

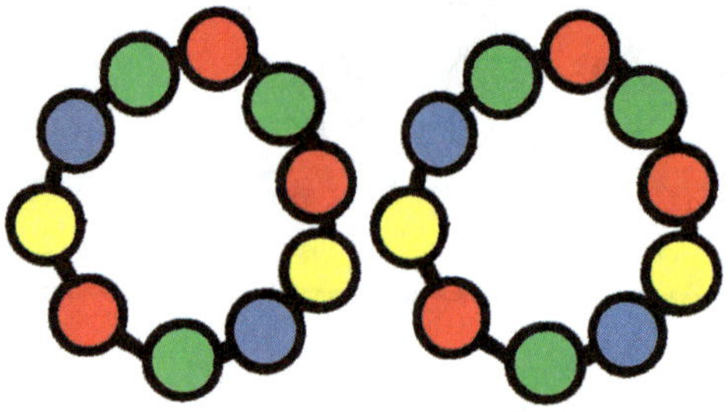

26

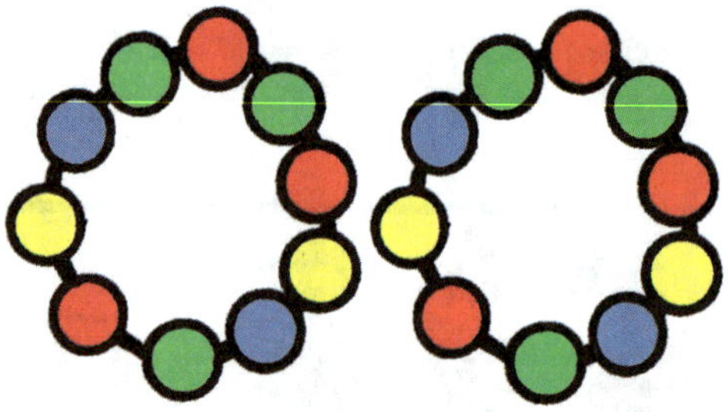

30

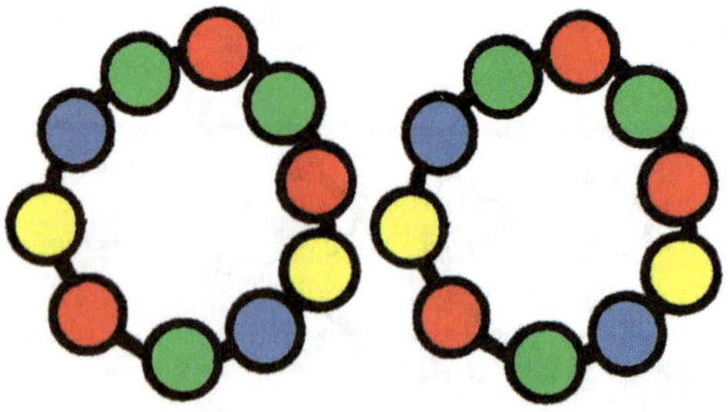

27

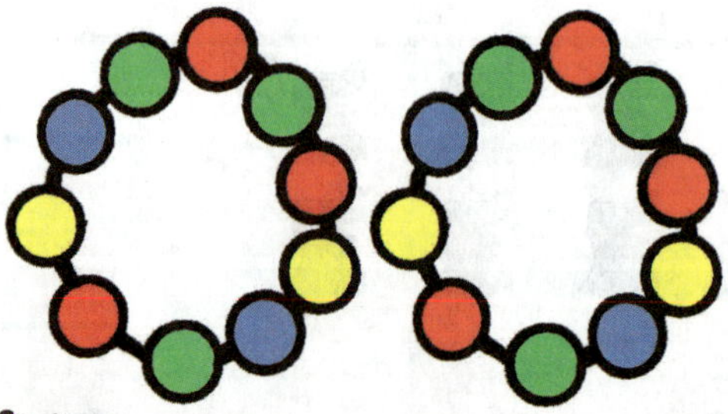

Directions: Have your child look at the numbers in each row. Ask your child to count the beads, then draw more beads to match the number in the row. Have your child practice writing each number in the space provided. ***Skill:*** Identifying groups of 26 to 30.

NAME ______________________________

REVIEW

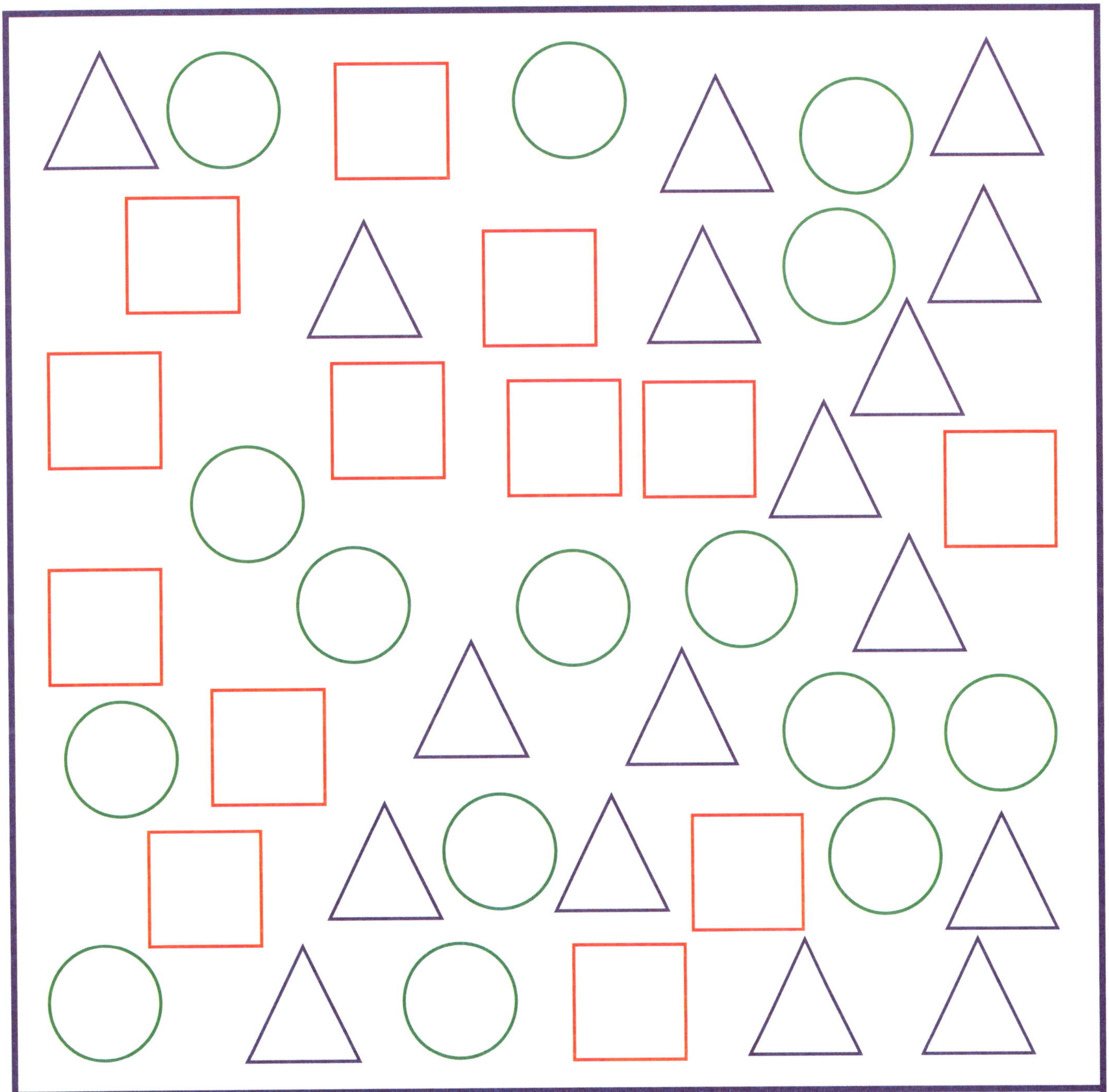

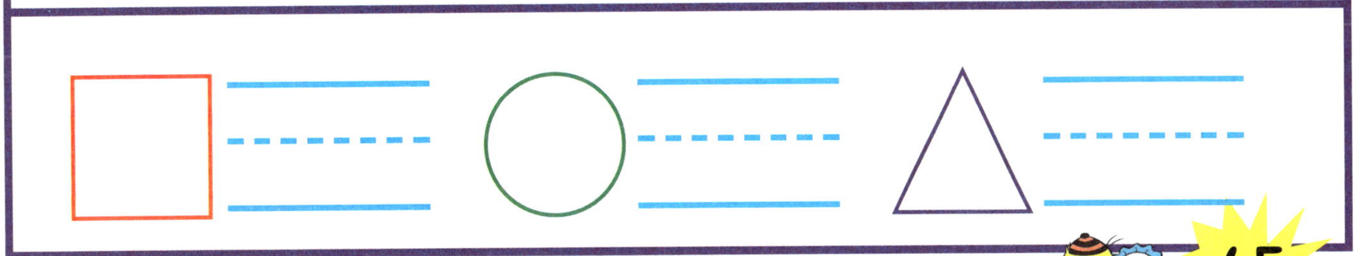

Directions: Have your child look at the objects at the bottom of the page. Ask your child to find, then count the objects in the picture. Have your child write how many of each object, then circle the greatest number. ***Skill:*** Counting; identifying more.

NAME ____________________

CALENDAR

JULY						
Sunday	Monday	Tuesday	Wednesday	Thursday	Friday	Saturday
				1	2	3
4	5	6	7	8	9	10
11	12	13	14	15	16	17
18	19	20	21	22	23	24
25	26	27	28	29	30	31

1. Trip to Pismo Beach on the first Wednesday. (RED)
2. Trick Elmer Fudd on the third Saturday. (PURPLE)
3. Daffy's party on the fourth Saturday. (GREEN)
4. Make carrot cake on the second Tuesday. (YELLOW)
5. Go to opera with Lola on the first Sunday. (BLUE)
6. Singing lessons each Monday. (ORANGE)
7. Tap dance lessons each Thursday. (BROWN)
8. Visit Mars on first Saturday. (BLACK)

Directions: Have your child study the calendar. Ask your child to color the dates, using the directions listed below the calendar as a reference. ***Skill:*** Reading a calendar.

ANSWER KEY

NAME

ONE AND TWO

Circle 1 . Place an X on 2 .

Place an X on 2 . Circle 1 .

Circle all but 2 .

Directions: Have your child circle or place an X on one or two birds according to the directions in each row.
Skill: Identifying one and two.

5

NAME

WRITING 1 AND 2

2 | 1
1 | 2
2 | 2

6

Directions: Have your child trace the numbers 1 and 2 at the top of the page. Ask your child to count the hats in each block and write 1 or 2 to tell how many. Then have your child color the hats.
Skill: Identifying and writing 1 and 2.

NAME

THREE AND FOUR

4 | 1
4 | 2
2 | 3

Directions: Have your child count, then draw the missing wheels in each block. **Skill:** Identifying groups of one to four.

7

NAME

WRITING 3 AND 4

4 3 3 4 2

8

Directions: Have your child count each kind of clothing, then write 3 or 4 to tell how many. **Skill:** Identifying and writing 3 and 4.

NAME

FIVE

5 | 3
2 | 5
4 | 5

Directions: Have your child look at the number in each block, then draw that number of objects.
Skill: Identifying groups of two to five.

9

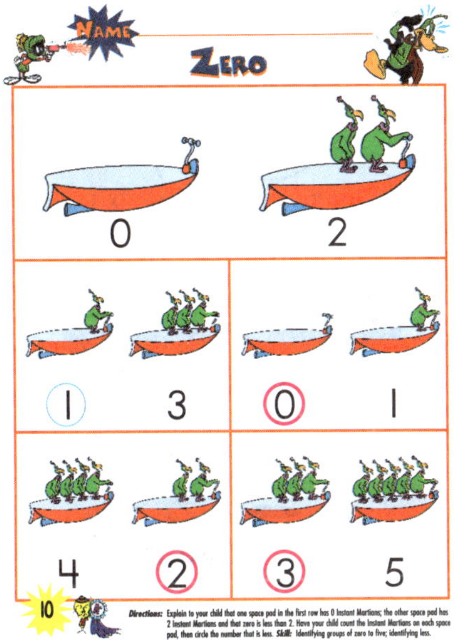
NAME

ZERO

0 2

1 3 | 0 1

4 2 | 3 5

10

Directions: Explain to your child that one space pod in the first row has 0 Instant Martians; the other space pod has 2 Instant Martians and that zero is less than 2. Have your child count the Instant Martians on each space pod, then circle the number that is less. **Skill:** Identifying groups of zero to five; identifying less.

Answer Key

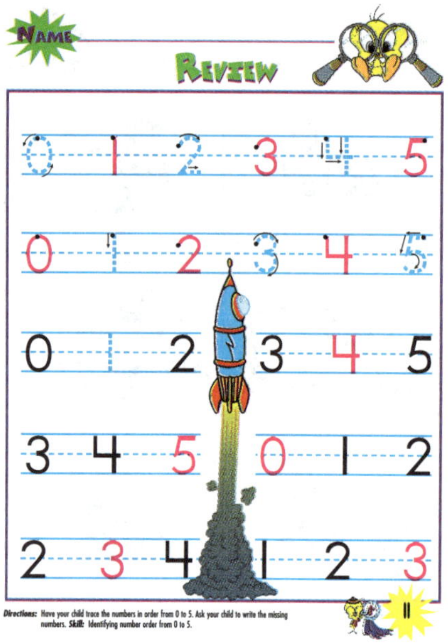
NAME
Review
0 1 2 3 4 5
0 1 2 3 4 5
0 1 2 3 4 5
3 4 5 0 1 2
2 3 4 1 2 3
Directions: Have your child trace the numbers in order from 0 to 5. Ask your child to write the missing numbers. Skill: Identifying number order from 0 to 5.
11

NAME
Problem Solving
Color the [shirt].
Color 3 BLUE. (b) Color 2 BROWN. (br)
Color 4 YELLOW. (y) Color 5 RED. (r)
Color 1 BLACK. (bl) Coloring will vary.
12
Directions: Have your child color the rabbits' shirts in the stadium following the color directions provided above the picture. Skill: Identifying groups of one to five.

NAME
Counting and Matching
4 5
1 2
Directions: Have your child identify the object in each block, count the same objects that appear in the picture, then write the number to tell how many. Ask your child to draw an X on the extra pictures. Skill: Counting and matching groups of one to five.
13

NAME
Ordinals to Fifth
14
Directions: Have your child circle the first character, put a triangle around the second character, put a box around the third character, underline the fourth character, and put an X on the fifth character. Skill: Identifying ordinal positions first to fifth.

NAME
Progress Check
1. 3 2. 1
3. 2 4. 0
5. 4 6. 5
Directions: Have your child find the animals that are hiding in the picture, then write the number to tell how many. Skill: Matching and counting.
15

NAME
Six
one 1
three 3
six 6
two 2
four 4
16
Directions: Have your child write the number that is written on each flowerpot, then draw flowers to show how many. Skill: Identifying and writing numbers from 1 to 6.

ANSWER KEY

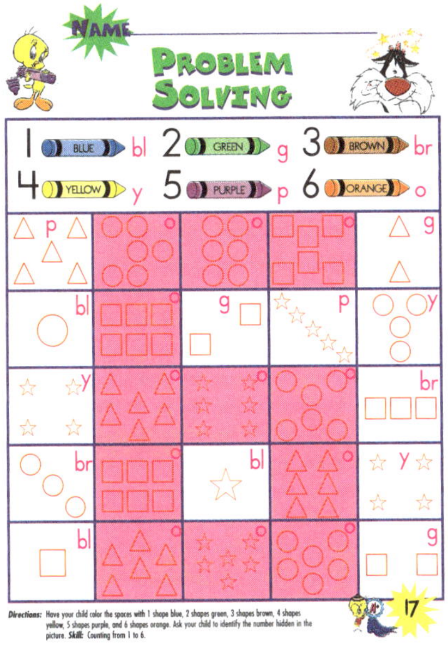
NAME
PROBLEM SOLVING
1 BLUE bl 2 GREEN g 3 BROWN br
4 YELLOW y 5 PURPLE p 6 ORANGE o
17
Directions: Have your child color the spaces with 1 shape blue, 2 shapes green, 3 shapes brown, 4 shapes yellow, 5 shapes purple, and 6 shapes orange. Ask your child to identify the number hidden in the picture. Skill: Counting from 1 to 6.

NAME
SEVEN
3 more carrots
CARROTS
2 more chairs and
2 more plates
18
Directions: Have your child draw carrots so that there are 7 in all in Elmer's garden. Then ask your child to set Elmer's table with enough chairs and plates for 7 people. Skill: Identifying seven.

NAME
EIGHT
19
Directions: Have your child draw windows on each of the objects so that there are 8 in all.
Skill: Identifying eight.

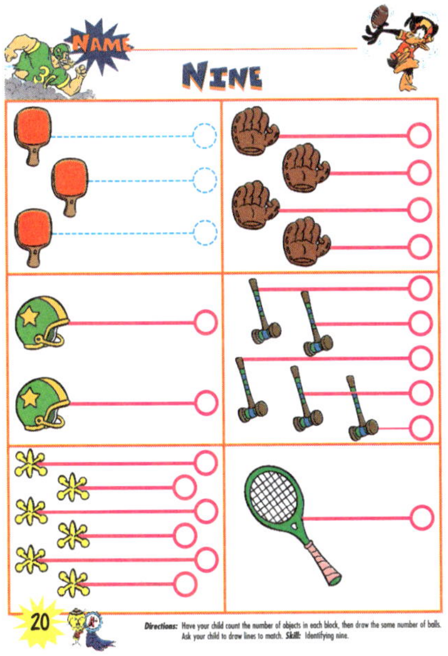
NAME
NINE
20
Directions: Have your child count the number of objects in each block, then draw the same number of balls. Ask your child to draw lines to match. Skill: Identifying nine.

NAME
REVIEW
7
9
1
5
3
8
2
4
6
21
Directions: Have your child finish coloring each cake. Ask your child to draw the missing candles.
Skill: Reviewing numbers from 1 to 9.

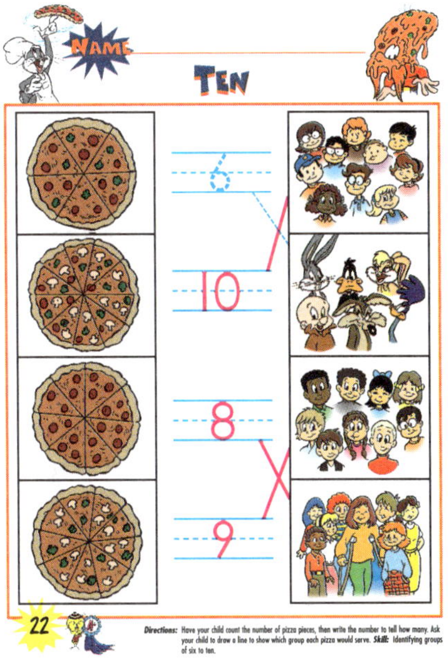
NAME
TEN
6
10
8
9
22
Directions: Have your child count the number of pizza pieces, then write the number to tell how many. Ask your child to draw a line to show which group each pizza would serve. Skill: Identifying groups of six to ten.

ANSWER KEY

NAME
GRAPHING
1 2 3 4 5 6 7 8 9 10
Directions: Have your child count the number of sea animals in the picture. Then ask your child to color in the graph to show how many. Skill: Using a graph.
23

NAME
PROGRESS CHECK
4 5 6 7
1 2 2 3
8 9 7 8
3 4 9 10
5 6 0 1
24
Directions: Have your child look at the numbers on each balloon, then write the number that comes just after. Skill: Identifying number order from 0 to 10.

NAME
MORE
Directions: Have your child count the Instant Martians in each row then following the example in row 1, ask your child to draw one more watering can than Martians in the second row, two more watering cans than Martians in the third row, and three more watering cans than Martians in the fourth row. Skill: Identifying more; reviewing ordinals.
25

NAME
ONE MORE, ONE LESS
2 3 4
4 5 6
8 9 10
1 2 3
6 7 8
3 4 5
26
Directions: Have your child look at the number that appears in the center of each row. Following the example in the first row, ask your child to write the number that is one less to the left and the number that is one more to the right. Skill: Identifying one more and one less; reviewing number order.

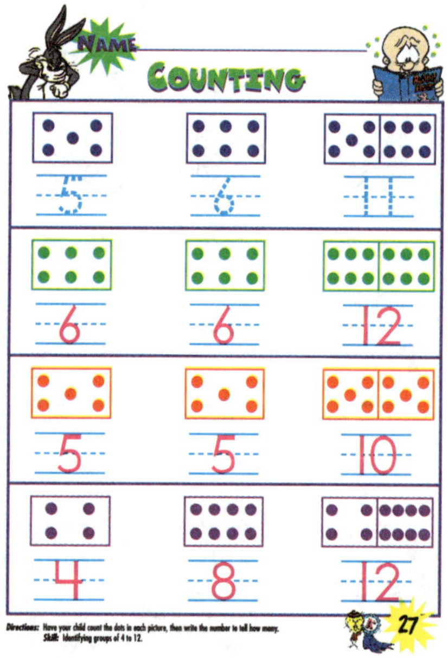
NAME
COUNTING
5 6 11
6 6 12
5 5 10
4 8 12
Directions: Have your child count the dots in each picture, then write the number to tell how many. Skill: Identifying groups of 4 to 12.
27

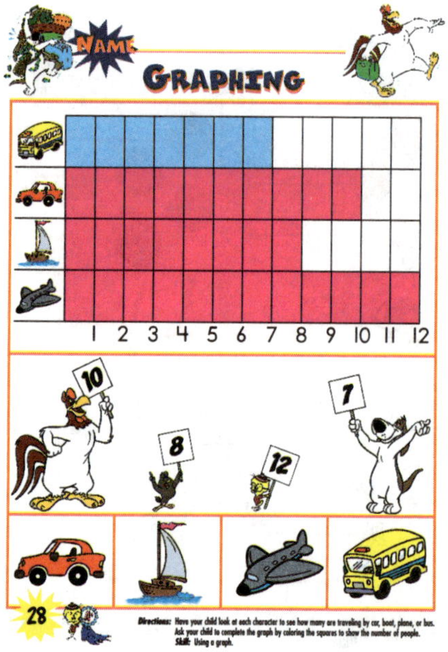
NAME
GRAPHING
1 2 3 4 5 6 7 8 9 10 11 12
10
8
12
7
28
Directions: Have your child look at each character to see how many are traveling by car, boat, plane, or bus. Ask your child to complete the graph by coloring the squares to show the number of people. Skill: Using a graph.

ANSWER KEY

NAME
BALL
Directions: Have your child place an X on the objects in the scene that have the same shape as a ball.
Skill: Identifying a ball.
29

NAME
CAN
Directions: Have your child place an X on the objects in the scene that have the same shape as a can.
Skill: Identifying a can.
30

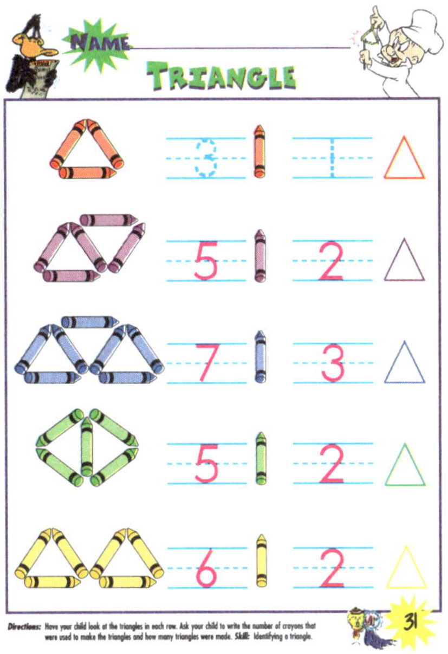
NAME
TRIANGLE
3 1
5 2
7 3
5 2
6 2
Directions: Have your child look at the triangles in each row. Ask your child to write the number of crayons that were used to make the triangles and how many triangles were made. Skill: Identifying a triangle.
31

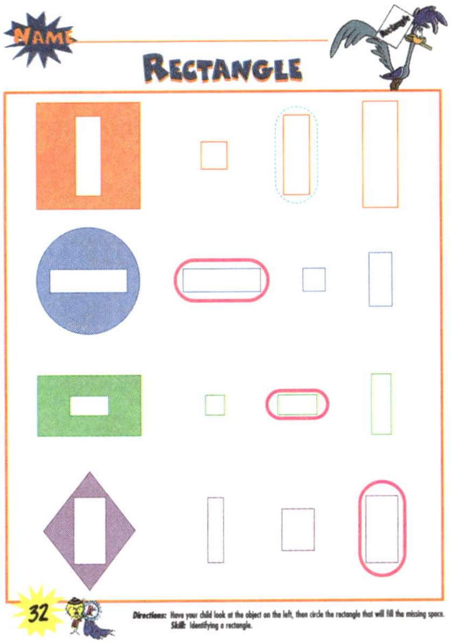
NAME
RECTANGLE
32
Directions: Have your child look at the object on the left, then circle the rectangle that will fill the missing space.
Skill: Identifying a rectangle.

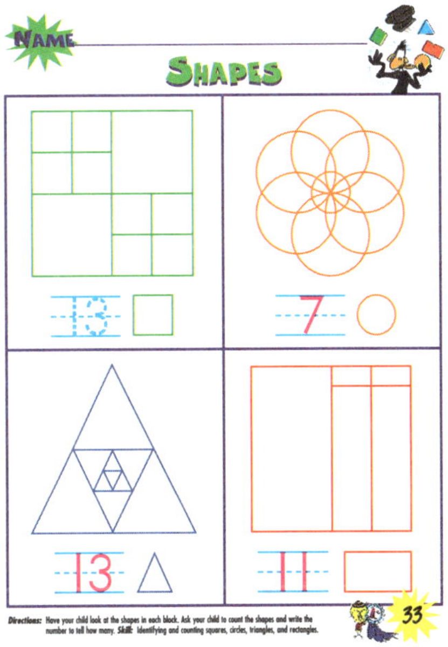
NAME
SHAPES
13
7
13
11
Directions: Have your child look at the shapes in each block. Ask your child to count the shapes and write the number to tell how many. Skill: Identifying and counting squares, circles, triangles, and rectangles.
33

NAME
REVIEW
RED
BLUE
YELLOW
GREEN
r
b
y
g
TOYS
34
Directions: Have your child look at the scene and color the circles red, the squares blue, the triangles yellow and the rectangles green. Skill: Reviewing shapes.

ANSWER KEY

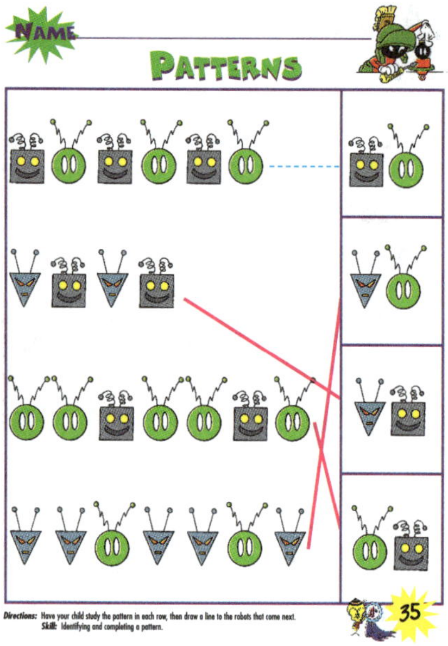
NAME
PATTERNS
35
Directions: Have your child study the pattern in each row, then draw a line to the robots that come next.
Skill: Identifying and completing a pattern.

NAME
ONE HALF
36
Directions: Have your child look at each object, then draw a line to make two equal parts. Ask your child to color one half of the object blue. Skill: Identifying one half.

NAME
SEQUENCE
MAIL
3
4
Dear Daffy, I hope you're getting enough to eat. Love, Mom
1
2
37
Directions: Have your child look at the pictures in each block, then write 2, 3, and 4 on the lines to show what happens second, third, and last. Skill: Identifying a sequence of events.

NAME
MORE TIME, LESS TIME
38
Directions: Have your child trace each path and make an X in the box to show which path took more time.
Skill: Estimating time.

NAME
TIME
4
10
7
2
3
6
39
Directions: Have your child look at the number beside each clock, then draw the little hand so it points to the number. Skill: Recognizing a clock face.

NAME
PENNY
6¢
9¢
7¢
10¢
40
Directions: Have your child circle the pennies needed to buy each item. Skill: Identifying and counting pennies.

Answer Key

Directions: Have your child count the pennies and nickels at the beginning of each row. Ask your child to circle the two items in each row that can be bought if all of the money is spent. ***Skill:*** Identifying and counting pennies and nickels.

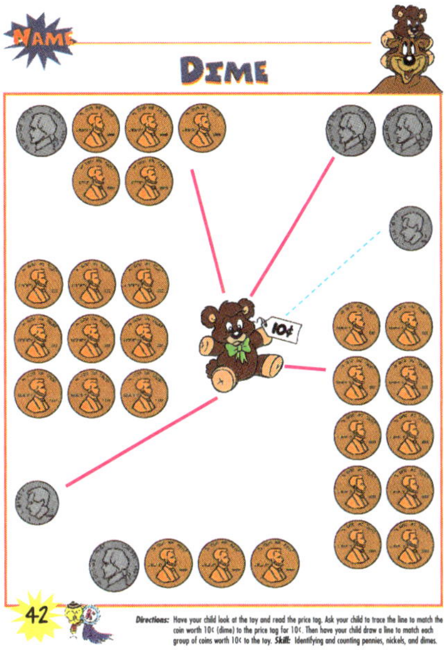

Directions: Have your child look at the toy and read the price tag. Ask your child to trace the line to match the coin worth 10¢ (dime) to the price tag for 10¢. Then have your child draw a line to match each group of coins worth 10¢ to the toy. ***Skill:*** Identifying and counting pennies, nickels, and dimes.

Directions: Have your child count the coins in each block, then write the amount on each piggy bank. ***Skill:*** Identifying and counting pennies, nickels, and dimes.

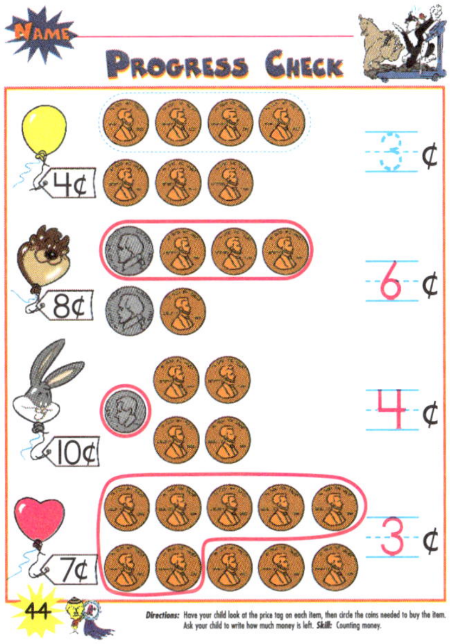

Directions: Have your child look at the price tag on each item, then circle the coins needed to buy the item. Ask your child to write how much money is left. ***Skill:*** Counting money.

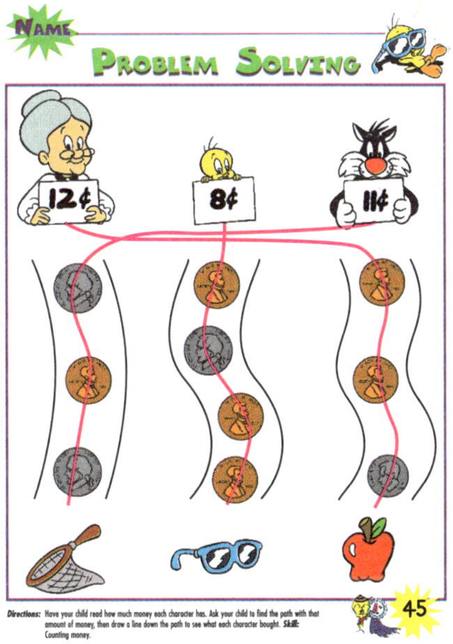

Directions: Have your child read how much money each character has. Ask your child to find the path with that amount of money, then draw a line down the path to see what each character bought. ***Skill:*** Counting money.

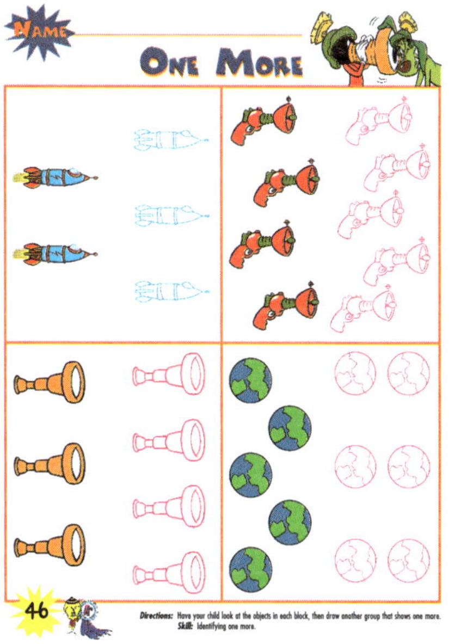

Directions: Have your child look at the objects in each block, then draw another group that shows one more. ***Skill:*** Identifying one more.

Answer Key

Name
Counting and Matching
5
3
4
6
Directions: Have your child look at the number at the end of each row that shows how many in all. Ask your child to draw more shapes to match the number. Skill: Counting and matching.
47

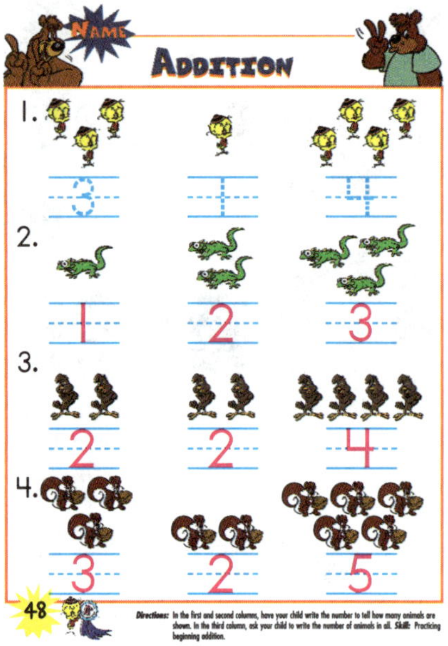
Name
Addition
1. 3 1 4
2. 1 2 3
3. 2 2 4
4. 3 2 5
48
Directions: In the first and second columns, have your child write the number to tell how many animals are shown. In the third column, ask your child to write the number of animals in all. Skill: Practicing beginning addition.

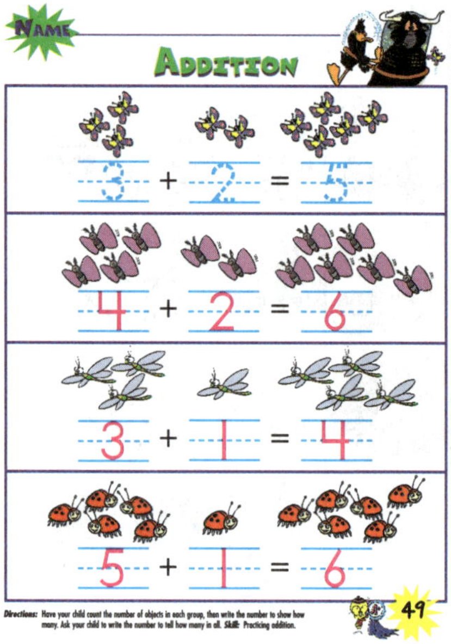
Name
Addition
3 + 2 = 5
4 + 2 = 6
3 + 1 = 4
5 + 1 = 6
Directions: Have your child count the number of objects in each group, then write the number to show how many. Ask your child to write the number to tell how many in all. Skill: Practicing addition.
49

Name
Review
2 + 2 = 4
1 + 1 = 2
1 + 2 = 3
1 + 3 = 4
50
Directions: Have your child find the groups of animals in the picture. Following the example, ask your child to write an addition sentence for each group. Skill: Writing addition sentences.

Name
One Less
Directions: Have your child look at the objects in each block, then draw another group that shows one less. Skill: Identifying one less.
51

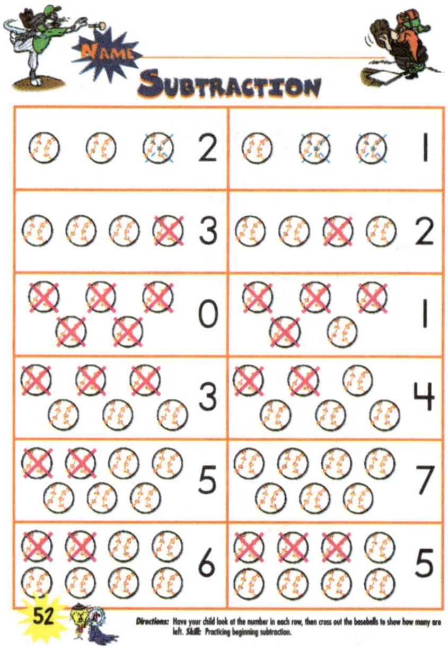
Name
Subtraction
2
1
3
2
0
1
3
4
5
7
6
5
52
Directions: Have your child look at the number in each row, then cross out the baseballs to show how many are left. Skill: Practicing beginning subtraction.

Answer Key

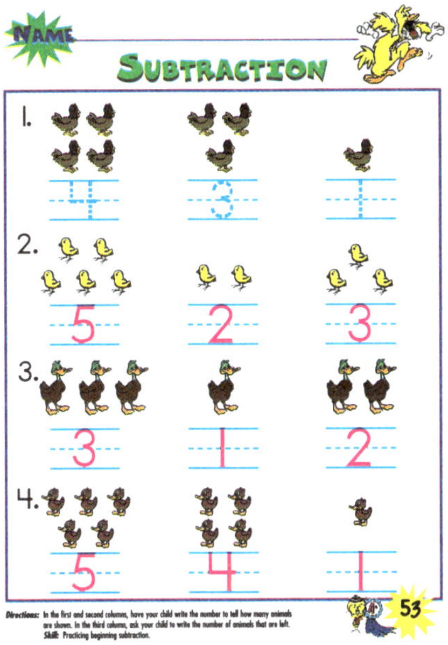
NAME
Subtraction
1. 4 3 1
2. 5 2 3
3. 3 1 2
4. 5 4 1
53

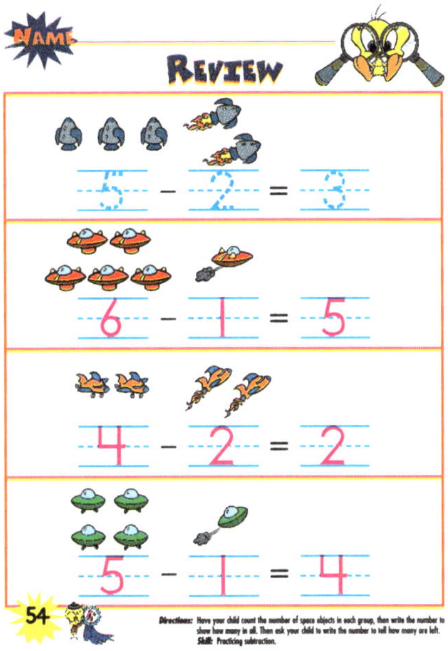
NAME
Review
5 – 2 = 3
6 – 1 = 5
4 – 2 = 2
5 – 1 = 4
54

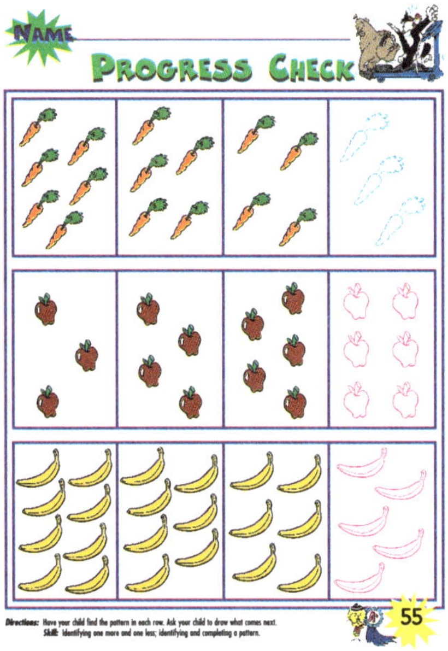
NAME
Progress Check
55

NAME
Problem Solving
1+2=3
3+1=4
2+3=5
56

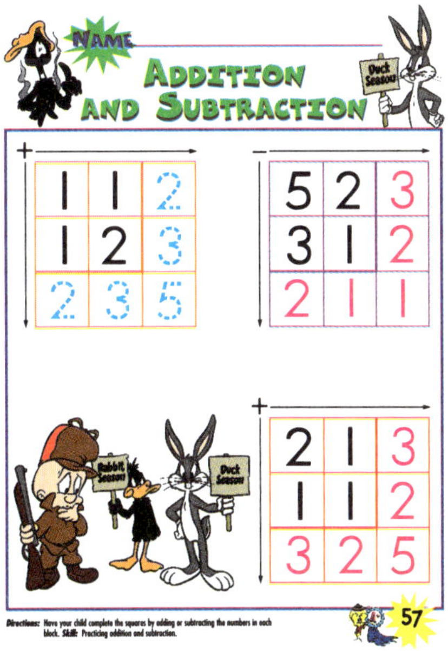
NAME
Addition and Subtraction
Duck Season
Rabbit Season
+
1 1 2
1 2 3
2 3 5
–
5 2 3
3 1 2
2 1 1
+
2 1 3
1 1 2
3 2 5
57

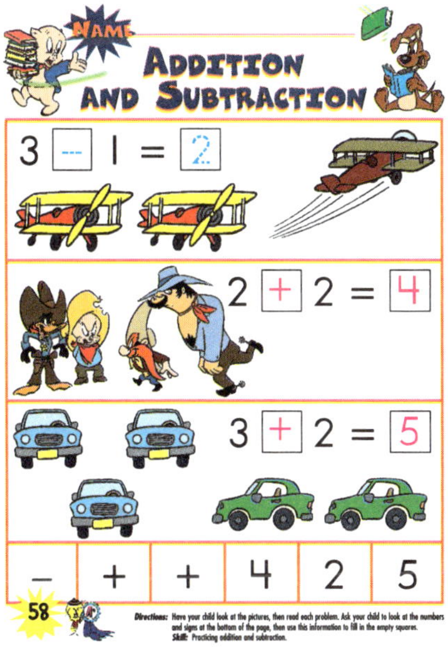
NAME
Addition and Subtraction
3 – 1 = 2
2 + 2 = 4
3 + 2 = 5
– + + 4 2 5
58

ANSWER KEY

NAME
COUNTING

4
6
5
7

Directions: Have your child circle the robots with square heads and place an X on the robots with round heads. Ask your child to count how many robots in all, then write the answer in the space provided. ***Skill:*** Counting and matching.

59

NAME
MORE THAN 10

60

Directions: Have your child look at the picture, then circle the objects at the bottom of the page that show more than 10. ***Skill:*** Identifying groups of 10 and more.

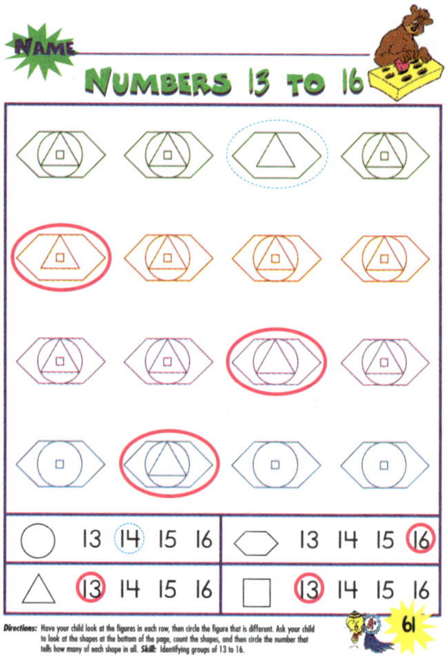
NAME
NUMBERS 13 TO 16

○	13	14	15	16	⬡	13	14	15	16
△	13	14	15	16	□	13	14	15	16

Directions: Have your child look at the figures in each row, then circle the figure that is different. Ask your child to look at the shapes at the bottom of the page, count the shapes, and then circle the number that tells how many of each shape in all. ***Skill:*** Identifying groups of 13 to 16.

61

NAME
NUMBERS 14 TO 20

15	17	20	19	14	18

20
14
15
17
17
18
19
15

62

Directions: Have your child study the chart at the top of the page. Ask your child to write how many of each bird, then circle the bird in each block that has the greater number. ***Skill:*** Identifying groups of 14 to 20; identifying more.

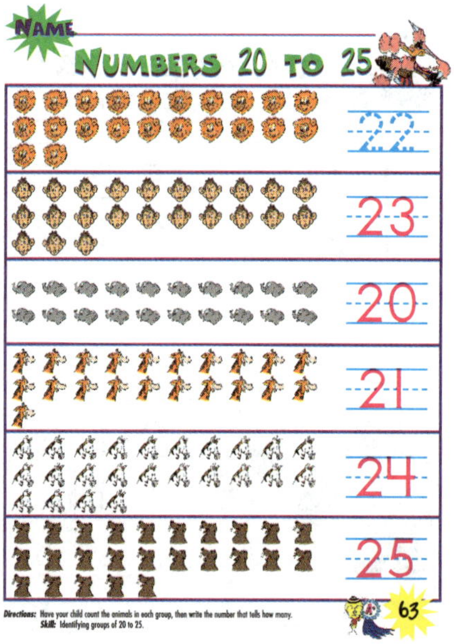
NAME
NUMBERS 20 TO 25

22
23
20
21
24
25

Directions: Have your child count the animals in each group, then write the number that tells how many. ***Skill:*** Identifying groups of 20 to 25.

63

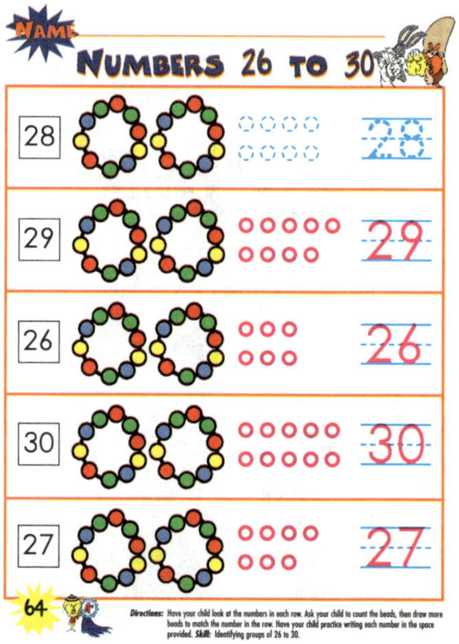
NAME
NUMBERS 26 TO 30

28 — 28
29 — 29
26 — 26
30 — 30
27 — 27

64

Directions: Have your child look at the numbers in each row. Ask your child to count the beads, then draw more beads to match the number in the row. Have your child practice writing each number in the space provided. ***Skill:*** Identifying groups of 26 to 30.

ANSWER KEY

NAME

REVIEW

13 15 17

Directions: Have your child look at the objects at the bottom of the page. Ask your child to find, then count the objects in the picture. Have your child write how many of each object, then circle the greatest number. Skill: Counting; identifying more.

65

NAME

CALENDAR

b=blue br=brown JULY o=orange r=red
bl=black g=green p=purple y=yellow

Sunday	Monday	Tuesday	Wednesday	Thursday	Friday	Saturday
				br 1	2	bl 3
b 4	o 5	6	r 7	br 8	9	10
11	o 12	y 13	14	br 15	16	p 17
18	o 19	20	21	br 22	23	g 24
25	o 26	27	28	br 29	30	31

1. Trip to Pismo Beach on the first Wednesday. RED
2. Trick Elmer Fudd on the third Saturday. PURPLE
3. Daffy's party on the fourth Saturday. GREEN
4. Make carrot cake on the second Tuesday. YELLOW
5. Go to opera with Lola on the first Sunday. BLUE
6. Singing lessons each Monday. ORANGE
7. Tap dance lessons each Thursday. BROWN
8. Visit Mars on first Saturday.

66

Directions: Have your child study the calendar. Ask your child to color the dates, using the directions listed below the calendar as a reference. Skill: Reading a calendar.

A McGraw•Hill/Warner Bros. Workbook

The McGraw-Hill Junior Academic Series

CERTIFICATE OF ACCOMPLISHMENT

THIS CERTIFIES THAT

____

HAS SUCCESSFULLY COMPLETED
THE JUNIOR ACADEMIC'S™

Grade K Enrichment Math

WORKBOOK.
CONGRATULATIONS AND THAT'S ALL FOLKS!

The McGraw·Hill Companies
PUBLISHER

Bugs Bunny
BUGS BUNNY, EDITOR-IN-CHIEF

RECEIVE THE McGRAW-HILL PARENT NEWSLETTER FREE!

Thank you for expressing interest in the successful education of your child. With the purchase of this workbook, we know that you are committed to your child's development and future success. We at ***McGraw-Hill Consumer Products*** *would like to help you make a difference in the education of your child by offering a quarterly newsletter that provides current topics on education and activities that you and your child can work on together.*

To receive a free copy of our newsletter, please provide us with the following information:

Name ________________ *Store where book purchased* ________________

Address ________________ *Grade Level of book purchased* ________________

City ________ *State* ____ *Zip* ____ *Title of book purchased* ________________

e-mail (if applicable) ________________

Mail to:
Parent Newsletter
c/o McGraw-Hill Consumer Products
251 Jefferson Street, M.S. #12
Waldoboro, ME 04572

Or Call 800-298-4119

Or visit us at:

www.MHkids.com

This offer is limited to residents of the United States and Canada and is only in effect for as long as the newsletter is published.
The information that you provide will not be given, rented, or sold to any company.

From the Nation's #1 Educational Publisher K-12

The McGraw-Hill Junior Academic Series

Kindergarten

Enrichment Reading

Challenging and Fun Activities

Vowels & Consonants

Capital & Lowercase Letters

Table of Contents

Table of Contents (continued)

NAME ______________________

MY BOOK

Directions: Have your child personalize this workbook by drawing a self-portrait and writing his or her name.

NAME

BEGINNING SOUND C

Cc

Directions: Have your child talk about the things in the picture and identify each object or animal, then circle the things that begin with the /k/ sound as in ***cupboard***. ***Skill:*** Identifying the beginning sound c.

Name ______________________________

Consonants: C

Cc

My word that starts with c:

Directions: Have your child look at the containers for words that begin with *C*, then put an X on those containers that have *C* words. Ask your child to practice writing *Cc*, then draw and label a picture for another word that begins with *C*. ***Skill:*** Identifying and writing Cc.

NAME

BEGINNING SOUND P

Pp

Directions: Have your child talk about the things in the picture and identify each object or animal, then circle the things that begin with the /p/ sound as in ***picture***. ***Skill:*** Identifying the beginning sound p.

NAME ______________________________

CONSONANTS: P

Pp

My word that starts with p:

Directions: Have your child look at the containers for words that begin with *P*, then put an X on those containers that have *P* words. Ask your child to practice writing *Pp*, then draw and label a picture for another word that begins with *P*. ***Skill:*** Identifying and writing Pp.

Name

Beginning Sound H

Hh

Directions: Have your child talk about the things in the picture and identify each object or animal, then circle the things that begin with the /h/ sound as in ***house***. ***Skill:*** Identifying the beginning sound h.

NAME

CONSONANTS: H

Hh

My word that starts with h:

Directions: Have your child look at the signs for words that begin with *H*, then put an X on those signs that have *H* words. Ask your child to practice writing *Hh*, then draw and label a picture for another word that begins with *H*. ***Skill:*** Identifying and writing Hh.

NAME ____________________

Beginning Sound M

Mm

Directions: Have your child talk about the things in the picture and identify each object or animal, then circle the things that begin with the /m/ sound as in ***mountain***. ***Skill:*** Identifying the beginning sound m.

NAME ______________________________

CONSONANTS: M

Mm

My word that starts with m:

Directions: Have your child look at the places on the map for words that begin with *M*, then circle the names of the places that begin with *M*. Ask your child to practice writing *Mm*, then draw and label a picture for another word that begins with *M*. **_Skill:_** Identifying and writing Mm.

NAME

BEGINNING SOUND S

Ss

Directions: Have your child talk about the things in the picture and identify each object or animal, then circle the things that begin with the /s/ sound as in ***seashore***. ***Skill:*** Identifying the beginning sound s.

NAME ____________________

CONSONANTS: S

Ss

My word that starts with s:

Directions: Have your child look at the signs for words that begin with *S*, then underline those words that begin with *S*. Ask your child to practice writing *Ss*, then draw and label a picture for another word that begins with *S*.
Skill: Identifying and writing Ss.

NAME

BEGINNING SOUND B

Bb

Directions: Have your child talk about the things in the picture and identify each object or animal, then circle the things that begin with the /b/ sound as in ***boat***. ***Skill:*** Identifying the beginning sound b.

NAME

CONSONANTS: B

Bb

My word that starts with b:

Directions: Have your child look at the signs for words that begin with *B*, then put an X on those signs that have *B* words. Ask your child to practice writing *Bb*, then draw and label a picture for another word that begins with *B*.
Skill: Identifying and writing Bb.

NAME

BEGINNING SOUND G

Gg

Directions: Have your child talk about the things in the picture and identify each object or animal, then circle the things that begin with the /g/ sound as in ***garden***. ***Skill:*** Identifying the beginning sound g.

NAME

CONSONANTS: G

Gg

My word that starts with g:

Directions: Have your child look at the signs for words that begin with *G*, then put an X on those signs that have *G* words. Ask your child to practice writing *Gg*, then draw and label a picture for another word that begins with *G*. ***Skill:*** Identifying and writing Gg.

NAME

BEGINNING SOUND SHORT A

Aa

Directions: Have your child talk about the things in the picture and identify each object or animal, then circle the things that begin with the /a/ sound as in ***adventure***. ***Skill:*** Identifying the beginning short a vowel sound.

NAME ______________________

VOWELS: SHORT A

A a

My word that starts with a:

Directions: Have your child look at the signs for words that begin with *A*, then underline those words that begin with *A*. Ask your child to practice writing *Aa*, then draw and label a picture for another word that begins with *A*.
Skill: Identifying and writing Aa.

NAME

BEGINNING SOUND R

Directions: Have your child talk about the things in the picture and identify each object or animal, then circle the things that begin with the /r/ sound as in ***roller rink***. ***Skill:*** Identifying the beginning sound r.

NAME

CONSONANTS: R

Rr

My word that starts with r:

Directions: Have your child look at the signs for words that begin with *R*, then underline those words that begin with *R*. Ask your child to practice writing *Rr*, then draw and label a picture for another word that begins with *R*.
Skill: Identifying and writing Rr.

NAME ____________________

BEGINNING SOUND SHORT E

Directions: Have your child talk about the things in the picture and identify each object or animal, then circle the things that begin with the /e/ sound as in ***elephant***. ***Skill:*** Identifying the beginning short e vowel sound.

NAME

CONSONANTS: T

Tt

My word that starts with t:

Directions: Have your child look at the signs for words that begin with *T*, then underline those words that begin with *T*. Ask your child to practice writing *Tt*, then draw and label a picture for another word that begins with *T*.
Skill: Identifying and writing Tt.

NAME ____________________

BEGINNING SOUND K

Directions: Have your child talk about the things in the picture and identify each object or animal, then circle the things that begin with the /k/ sound as in ***kingdom***. ***Skill:*** Identifying the beginning sound k.

NAME

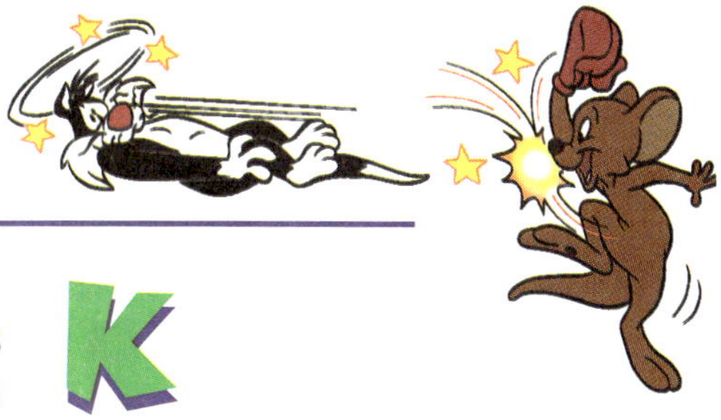

CONSONANTS: K

Kk

My word that starts with k:

Directions: Have your child look at the signs for words that begin with *K*, then underline those words that begin with *K*. Ask your child to practice writing *Kk*, then draw and label a picture for another word that begins with *K*.
Skill: Identifying and writing Kk.

NAME

BEGINNING SOUND F

Ff

Directions: Have your child talk about the things in the picture and identify each object or animal, then circle the things that begin with the /f/ sound as in ***farm***. ***Skill:*** Identifying the beginning sound f.

Name

Consonants: F

My word that starts with f:

Directions: Have your child look at the signs for words that begin with *F*, then underline those words that begin with *F*. Ask your child to practice writing *Ff*, then draw and label a picture for another word that begins with *F*.
Skill: Identifying and writing Ff.

NAME

BEGINNING SOUND Y

Directions: Have your child talk about the things in the picture and identify each object or animal, then circle the things that begin with the /y/ sound as in ***yard***. ***Skill:*** Identifying the beginning sound y.

NAME ____________________

CONSONANTS: Y

Yy

My word that starts with y:

Directions: Have your child look at the signs for words that begin with *Y*, then put an X on those signs that have *Y* words. Ask your child to practice writing *Yy*, then draw and label a picture for another word that begins with *Y*.
Skill: Identifying and writing Yy.

NAME

BEGINNING SOUND Z

Zz

Directions: Have your child talk about the things in the picture and identify each object or animal, then circle the things that begin with the /z/ sound as in ***zoo***. ***Skill:*** Identifying the beginning sound z.

NAME

CONSONANTS: Z

Z z

My word that starts with z:

Directions: Have your child look at the signs for words that begin with *Z*, then underline those words that begin with *Z*. Ask your child to practice writing *Zz*, then draw and label a picture for another word that begins with *Z*.
Skill: Identifying and writing Zz.

NAME ______________________________

BEGINNING SOUND Q

Directions: Have your child talk about the things in the picture and identify each object or animal, then circle the things that begin with the /kw/ sound as in ***quilt***. ***Skill:*** Identifying the beginning sound q.

NAME ______________________________

CONSONANTS: Q

Qq

My word that starts with q:

Directions: Have your child look at the signs for words that begin with *Q*, then underline those words that begin with *Q*. Ask your child to practice writing *Qq*, then draw and label a picture for another word that begins with *Q*.
Skill: Identifying and writing Qq.

Beginning Sound Short I

Directions: Have your child talk about the things in the picture and identify each object or animal, then circle the things that begin with the /i/ sound as in ***insect***. ***Skill:*** Identifying the beginning short i vowel sound.

NAME ______________________________

Vowels: Short I

I i

My word that starts with i:

Directions: Have your child look at the book covers for words that begin with *I*, then underline those words that begin with *I*. Ask your child to practice writing *Ii*, then draw and label a picture for another word that begins with *I*. ***Skill:*** Identifying and writing Ii.

NAME ______________________

FINAL SOUND X

Directions: Have your child talk about the things in the picture and identify each object or animal, then circle the things that end with the /ks/ sound as in ***fox***. ***Skill:*** Identifying the final sound x.

NAME

CONSONANTS: X

Xx

My word that starts with x:

Directions: Have your child look at the recipe for words that end with *X*, then underline those words that end with *X*. Ask your child to practice writing *Xx*, then draw and label a picture for another word that ends with *X*.
Skill: Identifying and writing Xx.

NAME ______________________________

Beginning Sound D

Dd

Directions: Have your child talk about the things in the picture and identify each object or animal, then circle the things that begin with the /d/ sound as in ***dig***. ***Skill:*** Identifying the beginning sound d.

NAME ______________________________

CONSONANTS: D

Dd

My word that starts with d:

Directions: Have your child look at the signs for words that begin with *D*, then underline those words that begin with *D*. Ask your child to practice writing *Dd*, then draw and label a picture for another word that begins with *D*.
Skill: Identifying and writing Dd.

NAME

Beginning Sound W

Directions: Have your child talk about the things in the picture and identify each object or animal, then circle the things that begin with the /w/ sound as in ***water***. ***Skill:*** Identifying the beginning sound w.

NAME

CONSONANTS: W

Ww

My word that starts with w:

Directions: Have your child look at the signs for words that begin with *W*, then underline those words that begin with *W*. Ask your child to practice writing *Ww*, then draw and label a picture for another word that begins with *W*.
Skill: Identifying and writing Ww.

Name ______________________

Beginning Sound J

Directions: Have your child talk about the things in the picture and identify each object or animal, then circle the things that begin with the /j/ sound as in ***jug***. ***Skill:*** Identifying the beginning sound j.

NAME

CONSONANTS: J

My word that starts with j:

Directions: Have your child look at the signs for words that begin with *J*, then underline those words that begin with *J*. Ask your child to practice writing *Jj*, then draw and label a picture for another word that begins with *J*.
Skill: Identifying and writing Jj.

NAME

BEGINNING SOUND L

Directions: Have your child talk about the things in the picture and identify each object or animal, then circle the things that begin with the /l/ sound as in ***line***. ***Skill:*** Identifying the beginning sound l.

NAME

CONSONANTS: L

My word that starts with l:

Directions: Have your child look at the labels for words that begin with *L*, then put an X on those labels that have *L* words. Ask your child to practice writing *Ll*, then draw and label a picture for another word that begins with *L*. ***Skill:*** Identifying and writing Ll.

Name ______________________________

Beginning Sound N

Directions: Have your child talk about the things in the picture and identify each object or animal, then circle the things that begin with the /n/ sound as in ***news***. ***Skill:*** Identifying the beginning sound n.

NAME ______________________

CONSONANTS: N

Nn

My word that starts with n:

Directions: Have your child look at the newspapers for words that begin with *N*, then underline those words that begin with *N*. Ask your child to practice writing *Nn*, then draw and label a picture for another word that begins with *N*. ***Skill:*** Identifying and writing Nn.

NAME

BEGINNING SOUND SHORT U

Directions: Have your child talk about the things in the picture and identify each object or animal, then circle the things that begin with the /u/ sound as in ***umbrella***. ***Skill:*** Identifying the beginning short u vowel sound.

NAME

Vowels: Short U

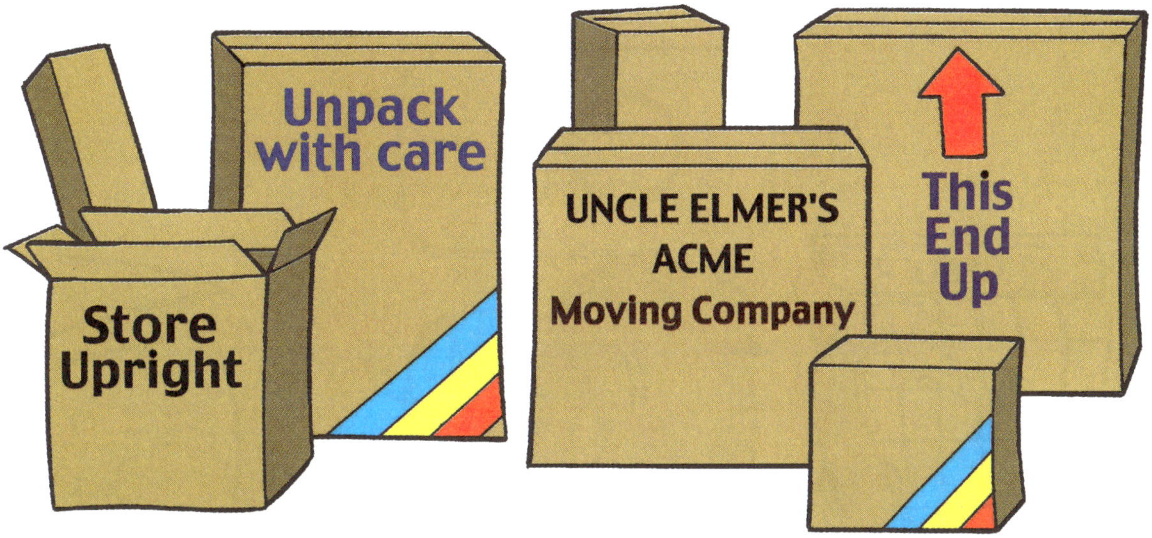

Uu

My word that starts with u:

Directions: Have your child look at the boxes for words that begin with *U*, then underline those words that begin with *U*. Ask your child to practice writing *Uu*, then draw and label a picture for another word that begins with *U*. ***Skill:*** Identifying and writing Uu.

NAME

BEGINNING SOUND SHORT O

Directions: Have your child talk about the things in the picture and identify each object or animal, then circle the things that begin with the /o/ sound as in ***octopus***. ***Skill:*** Identifying the beginning short o vowel sound.

NAME ______________________________

Vowels: Short O

Oo

My word that starts with o:

Directions: Have your child look at the calendar for words that begin with *O*, then put an X on the days that have *O* words. Ask your child to practice writing *Oo*, then draw and label a picture for another word that begins with *O*. ***Skill:*** Identifying and writing Oo.

NAME ______

BEGINNING SOUND V

Directions: Have your child talk about the things in the picture and identify each object or animal, then circle the things that begin with the /v/ sound as in ***vegetables***. ***Skill:*** Identifying the beginning sound v.

NAME

CONSONANTS: V

My word that starts with v:

Directions: Have your child look at the signs for words that begin with *V*, then underline those words that begin with *V*. Ask your child to practice writing *Vv*, then draw and label a picture for another word that begins with *V*.
Skill: Identifying and writing Vv.

NAME

PROGRESS CHECK

Dd

Directions: Have your child write the uppercase and lowercase letter for the first picture in each row. Then ask your child to circle the picture in that row that begins with the same sound. ***Skill:*** Writing uppercase and lowercase letters; identifying beginning sounds.

NAME

PROGRESS CHECK

Qq

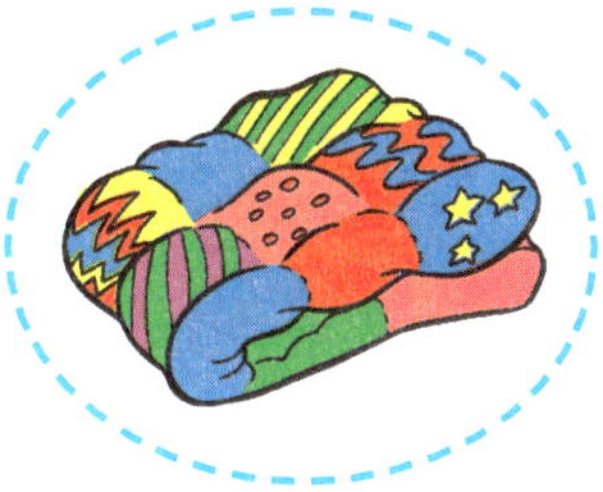

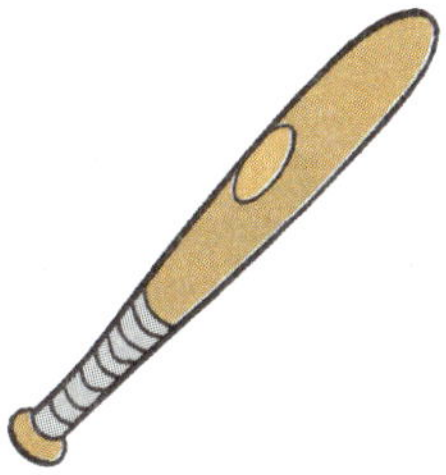

Directions: Have your child write the uppercase and lowercase letter for the first picture in each row. Then ask your child to circle the picture in that row that begins with the same sound. ***Skill:*** Writing uppercase and lowercase letters; identifying beginning sounds.

NAME

PROGRESS CHECK

Directions: Have your child write the uppercase and lowercase letter for the first picture in each row. Then ask your child to circle the picture in that row that begins with the same sound. ***Skill:*** Writing uppercase and lowercase letters; identifying beginning sounds.

NAME

PROGRESS CHECK

6

Directions: Have your child write the uppercase and lowercase letter for the first picture in each row. Then ask your child to circle the picture in that row that begins with the same sound. ***Skill:*** Writing uppercase and lowercase letters; identifying beginning sounds.

NAME

A Final Word

a

!

Directions: Have your child identify the first letter in each picture word, then unscramble the letters to find the secret message. ***Skill:*** Reviewing beginning sounds and letters.

ANSWER KEY

NAME

MY BOOK

Drawings will vary.

Directions: Have your child personalize this workbook by drawing a self-portrait and writing his or her name.

81

NAME

BEGINNING SOUND C

cups
cookie jar
cans

Cc

cat
candle
candlestick

82

Directions: Have your child talk about the things in the picture and identify each object or animal, then circle the things that begin with the /k/ sound as in *cupboard*. **Skill:** Identifying the beginning sound c.

NAME

CONSONANTS: C

Cc Cc Cc Cc

My word that starts with c:

Answers will vary.

Directions: Have your child look at the containers for words that begin with *C*, then put an X on those containers that have *C* words. Ask your child to practice writing *Cc*, then draw and label a picture for another word that begins with *C*. **Skill:** Identifying and writing *Cc*.

83

NAME

BEGINNING SOUND P

penguin
parrot

Pp

pig
paint brushes

84

Directions: Have your child talk about the things in the picture and identify each object or animal, then circle the things that begin with the /p/ sound as in *picture*. **Skill:** Identifying the beginning sound p.

NAME

CONSONANTS: P

Pp Pp Pp Pp

My word that starts with p:

Answers will vary.

Directions: Have your child look at the containers for words that begin with *P*, then put an X on those containers that have *P* words. Ask your child to practice writing *Pp*, then draw and label a picture for another word that begins with *P*. **Skill:** Identifying and writing Pp.

85

NAME

BEGINNING SOUND H

horse
hose
hive

Hh

hat
horn

86

Directions: Have your child talk about the things in the picture and identify each object or animal, then circle the things that begin with the /h/ sound as in *house*. **Skill:** Identifying the beginning sound h.

Answer Key

NAME
CONSONANTS: H
Elmer's Holly Wood Home
Elmer Fudd
RING BELL
HOME SWEET HOME
HOLD HANDRAIL
Hh Hh Hh Hh
My word that starts with h:
Answers will vary.
Directions: Have your child look at the signs for words that begin with H, then put an X on those signs that have H words. Ask your child to practice writing Hh, then draw and label a picture for another word that begins with H. Skill: Identifying and writing Hh.
87

NAME
BEGINNING SOUND M
moose
mittens
Mm
monkey
mouse
88
Directions: Have your child talk about the things in the picture and identify each object or animal, then circle the things that begin with the /m/ sound as in mountain. Skill: Identifying the beginning sound m.

NAME
CONSONANTS: M
MAP
MOOSE MOUNTAIN
MUFFIN SHOP
SKI LODGE
BUNNY SLOPE
MAGIC MOUNTAIN
CHAIR LIFT
PARKING LOT
Mm Mm Mm Mm
My word that starts with m:
Answers will vary.
Directions: Have your child look at the places on the map for words that begin with M, then circle the names of the places that begin with M. Ask your child to practice writing Mm, then draw and label a picture for another word that begins with M. Skill: Identifying and writing Mm.
89

NAME
BEGINNING SOUND S
sun
seal
sailboat
Ss
sailor
shells
sandcastle
LIFEGUARD
Duck on Duty
90
Directions: Have your child talk about the things in the picture and identify each object or animal, then circle the things that begin with the /s/ sound as in seashore. Skill: Identifying the beginning sound s.

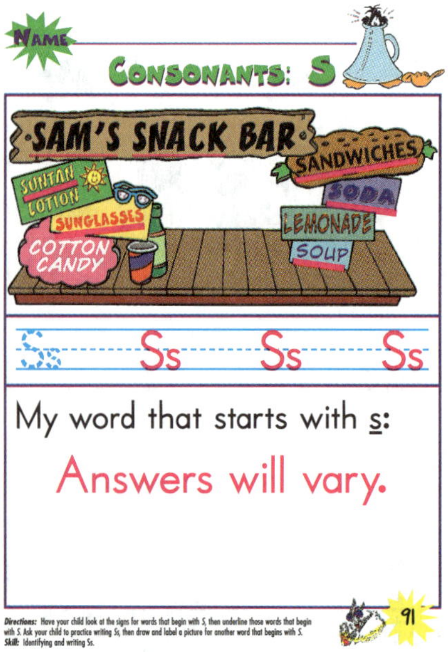
NAME
CONSONANTS: S
SAM'S SNACK BAR
SANDWICHES
SUNTAN LOTION
SUNGLASSES
SODA
LEMONADE
SOUP
COTTON CANDY
Ss Ss Ss Ss
My word that starts with s:
Answers will vary.
Directions: Have your child look at the signs for words that begin with S, then underline those words that begin with S. Ask your child to practice writing Ss, then draw and label a picture for another word that begins with S. Skill: Identifying and writing Ss.
91

NAME
BEGINNING SOUND B
buffalo
bear
butterfly
Bb
bib
basket
SAM'S
NOAH'S ARK
92
Directions: Have your child talk about the things in the picture and identify each object or animal, then circle the things that begin with the /b/ sound as in boat. Skill: Identifying the beginning sound b.

ANSWER KEY

NAME
CONSONANTS: B
My word that starts with b:
Answers will vary.
93
Directions: Have your child look at the signs for words that begin with *B*, then put an X on those signs that have *B* words. Ask your child to practice writing *Bb*, then draw and label a picture for another word that begins with *B*.
Skill: Identifying and writing Bb.

NAME
BEGINNING SOUND G
girl
gopher
goat
Gg
goose
glasses
grasshopper
94
Directions: Have your child talk about the things in the picture and identify each object or animal, then circle the things that begin with the /g/ sound as in ***garden***. **Skill:** Identifying the beginning sound g.

NAME
CONSONANTS: G
Garlic
Beans
Beets
Hollyhock
Gooseberry
Tulips
Gourds
Carrots
My word that starts with g:
Answers will vary.
95
Directions: Have your child look at the signs for words that begin with *G*, then put an X on those signs that have *G* words. Ask your child to practice writing *Gg*, then draw and label a picture for another word that begins with *G*. **Skill:** Identifying and writing Gg.

NAME
BEGINNING SOUND SHORT A
alligators
arrow
apples
Aa
ants
Alligators sign
ALLIGATOR ADVENTURE PARK
96
Directions: Have your child talk about the things in the picture and identify each object or animal, then circle the things that begin with the /a/ sound as in ***adventure***. **Skill:** Identifying the beginning short a vowel sound.

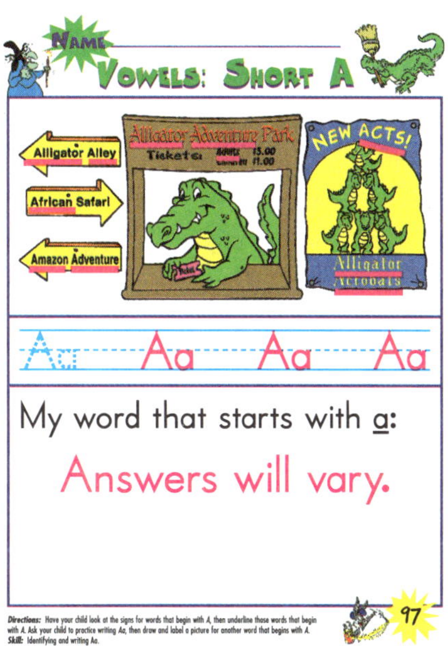
NAME
VOWELS: SHORT A
Alligator Alley
African Safari
Amazon Adventure
Alligator Adventure Park
Tickets
NEW ACTS!
Alligator Acrobats
My word that starts with a:
Answers will vary.
97
Directions: Have your child look at the signs for words that begin with *A*, then underline those words that begin with *A*. Ask your child to practice writing *Aa*, then draw and label a picture for another word that begins with *A*.
Skill: Identifying and writing Aa.

NAME
BEGINNING SOUND R
raccoons
rabbit
radio
Rr
rose
roller skates
98
Directions: Have your child talk about the things in the picture and identify each object or animal, then circle the things that begin with the /r/ sound as in ***roller rink***. **Skill:** Identifying the beginning sound r.

Answer Key

NAME

Consonants: R

RESTROOMS

SKATE RENTAL

RETURN SKATES HERE

SKATE SAFELY

Rr Rr Rr Rr

My word that starts with r:

Answers will vary.

Directions: Have your child look at the signs for words that begin with *R*, then underline those words that begin with *R*. Ask your child to practice writing *Rr*, then draw and label a picture for another word that begins with *R*. **Skill:** Identifying and writing Rr.

99

NAME

ELMER'S EXPRESS

Beginning Sound Short E

escalator
elevator
elephant

Ee

eggbeater
eggplant

SALE

100

Directions: Have your child talk about the things in the picture and identify each object or animal, then circle the things that begin with the /e/ sound as in ***elephant***. **Skill:** Identifying the beginning short e vowel sound.

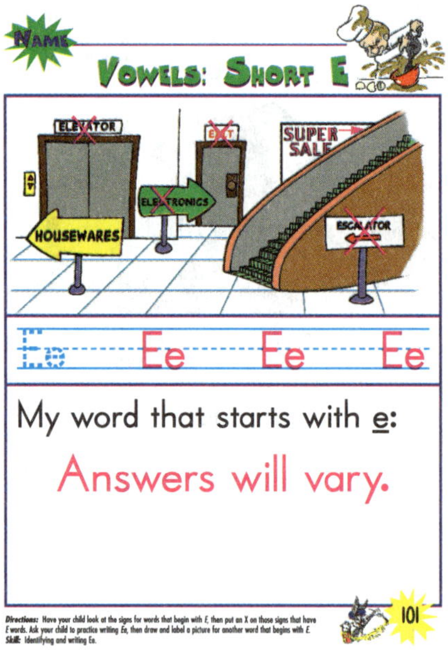

NAME

Vowels: Short E

ELEVATOR

EXIT

SUPER SALE

ELECTRONICS

HOUSEWARES

ESCALATOR

Ee Ee Ee Ee

My word that starts with e:

Answers will vary.

Directions: Have your child look at the signs for words that begin with *E*, then put an X on those signs that have *E* words. Ask your child to practice writing *Ee*, then draw and label a picture for another word that begins with *E*. **Skill:** Identifying and writing Ee.

101

NAME

Progress Check

alligator	egg	apple
buffalo	hippopotamus	bear
cat	cupboard	boat
elephant	house	elevator
goat	cow	garden

102

Directions: Have your child name the first picture in each row and then circle the picture for a word that begins with the same sound. **Skill:** Identifying beginning sounds.

NAME

Progress Check

horse	house	barn
monkey	giraffe	moose
panda	rhinoceros	pig
raccoon	rabbit	duck
seal	ball	sun

Directions: Have your child name the first picture in each row and then circle the picture for a word that begins with the same sound. **Skill:** Identifying beginning sounds.

103

NAME

Progress Check

car	Cc	panda	Pp
horse	Hh	monkey	Mm
seal	Ss	cat	Cc
mouse	Mm	sunglasses	Ss
penguin	Pp	hippopotamus	Hh

104

Directions: Have your child identify the first letter in each picture word and then write it as an uppercase and a lowercase letter. **Skill:** Identifying beginning letters; writing uppercase and lowercase letters.

Answer Key

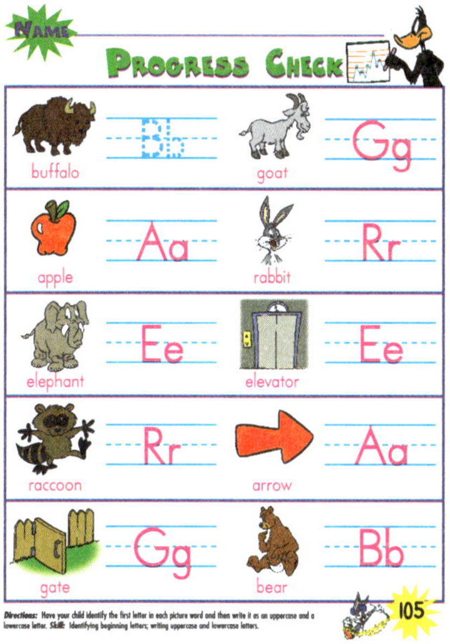
NAME
Progress Check
buffalo Bb
goat Gg
apple Aa
rabbit Rr
elephant Ee
elevator Ee
raccoon Rr
arrow Aa
gate Gg
bear Bb
Directions: Have your child identify the first letter in each picture word and then write it as an uppercase and a lowercase letter. ***Skill:*** Identifying beginning letters; writing uppercase and lowercase letters.
105

NAME
Beginning Sound T
Tt
turtle
tennis racquet
ties
tickets

106
Directions: Have your child talk about the things in the picture and identify each object or animal, then circle the things that begin with the /t/ sound as in ***train***. ***Skill:*** Identifying the beginning sound t.

NAME
Consonants: T

Tt Tt Tt Tt
My word that starts with t:
Answers will vary.
Directions: Have your child look at the signs for words that begin with *T*, then underline those words that begin with *T*. Ask your child to practice writing *Tt*, then draw and label a picture for another word that begins with *T*. ***Skill:*** Identifying and writing Tt.
107

NAME
Beginning Sound K
Kk
kites
keys
kangaroo
kaleidoscope
king
108
Directions: Have your child talk about the things in the picture and identify each object or animal, then circle the things that begin with the /k/ sound as in ***kingdom***. ***Skill:*** Identifying the beginning sound k.

NAME
Consonants: K

Kk Kk Kk Kk
My word that starts with k:
Answers will vary.
Directions: Have your child look at the signs for words that begin with *K*, then underline those words that begin with *K*. Ask your child to practice writing *Kk*, then draw and label a picture for another word that begins with *K*. ***Skill:*** Identifying and writing Kk.
109

NAME
Beginning Sound F
Ff
fence
fish
fox
farm stand
flowers
feathers

110
Directions: Have your child talk about the things in the picture and identify each object or animal, then circle the things that begin with the /f/ sound as in ***farm***. ***Skill:*** Identifying the beginning sound f.

ANSWER KEY

NAME

CONSONANTS: F

FARM STAND

My word that starts with f:

Answers will vary.

Directions: Have your child look at the signs for words that begin with *F*, then underline those words that begin with *F*. Ask your child to practice writing *Ff*, then draw and label a picture for another word that begins with *F*. ***Skill:*** Identifying and writing Ff.

111

NAME

BEGINNING SOUND Y

yo-yo
yarn

Yy

yak
yawn

YARD SALE

112

Directions: Have your child talk about the things in the picture and identify each object or animal, then circle the things that begin with the /y/ sound as in ***yard***. ***Skill:*** Identifying the beginning sound y.

NAME

CONSONANTS: Y

YARD SALE

Soft Ice Cream

My word that starts with y:

Answers will vary.

Directions: Have your child look at the signs for words that begin with *Y*, then put an X on those signs that have *Y* words. Ask your child to practice writing *Yy*, then draw and label a picture for another word that begins with *Y*. ***Skill:*** Identifying and writing Yy.

113

NAME

BEGINNING SOUND Z

zigzag
zebras

Zz

zookeeper
zipper

114

Directions: Have your child talk about the things in the picture and identify each object or animal, then circle the things that begin with the /z/ sound as in ***zoo***. ***Skill:*** Identifying the beginning sound z.

NAME

CONSONANTS: Z

LOONEY TUNES ZOO

ZOO CLOSED

My word that starts with z:

Answers will vary.

Directions: Have your child look at the signs for words that begin with *Z*, then underline those words that begin with *Z*. Ask your child to practice writing *Zz*, then draw and label a picture for another word that begins with *Z*. ***Skill:*** Identifying and writing Zz.

115

NAME

BEGINNING SOUND Q

queen
quail

Qq

quill
quilts

QUIET!

116

Directions: Have your child talk about the things in the picture and identify each object or animal, then circle the things that begin with the /kw/ sound as in ***quilt***. ***Skill:*** Identifying the beginning sound q.

ANSWER KEY

NAME

CONSONANTS: Q

Qq Qq Qq Qq

My word that starts with q:

Answers will vary.

Directions: Have your child look at the signs for words that begin with *Q*, then underline those words that begin with *Q*. Ask your child to practice writing *Qq*, then draw and label a picture for another word that begins with *Q*. ***Skill:*** Identifying and writing Qq.

117

NAME

BEGINNING SOUND SHORT I

inchworm
iguana

Ii

ink
inch-ruler

118

Directions: Have your child talk about the things in the picture and identify each object or animal, then circle the things that begin with the /i/ sound as in ***insect***. ***Skill:*** Identifying the beginning short i vowel sound.

NAME

VOWELS: SHORT I

Ii Ii Ii Ii Ii Ii

My word that starts with i:

Answers will vary.

Directions: Have your child look at the book covers for words that begin with *I*, then underline those words that begin with *I*. Ask your child to practice writing *Ii*, then draw and label a picture for another word that begins with *I*. ***Skill:*** Identifying and writing Ii.

119

NAME

FINAL SOUND X

box
mix

Xx

six

120

Directions: Have your child talk about the things in the picture and identify each object or animal, then circle the things that end with the /ks/ sound as in ***fox***. ***Skill:*** Identifying the final sound x.

NAME

CONSONANTS: X

Xx Xx Xx Xx

My word that starts with x:

Answers will vary.

Directions: Have your child look at the recipe for words that end with *X*, then underline those words that end with *X*. Ask your child to practice writing *Xx*, then draw and label a picture for another word that ends with *X*. ***Skill:*** Identifying and writing Xx.

121

NAME

BEGINNING SOUND D

dinosaurs
dump truck
donkey

Dd

dog
duck

122

Directions: Have your child talk about the things in the picture and identify each object or animal, then circle the things that begin with the /d/ sound as in ***dig***. ***Skill:*** Identifying the beginning sound d.

Answer Key

Consonants: D

Name

Dd Dd Dd Dd

My word that starts with d:

Answers will vary.

Directions: Have your child look at the signs for words that begin with *D*, then underline those words that begin with *D*. Ask your child to practice writing *Dd*, then draw and label a picture for another word that begins with *D*.
Skill: Identifying and writing Dd.

123

Beginning Sound W

Name

124

Directions: Have your child talk about the things in the picture and identify each object or animal, then circle the things that begin with the /w/ sound as in ***water***. **Skill:** Identifying the beginning sound w.

Consonants: W

Name

Ww Ww Ww Ww

My word that starts with w:

Answers will vary.

Directions: Have your child look at the signs for words that begin with *W*, then underline those words that begin with *W*. Ask your child to practice writing *Ww*, then draw and label a picture for another word that begins with *W*.
Skill: Identifying and writing Ww.

125

Beginning Sound J

Name

126

Directions: Have your child talk about the things in the picture and identify each object or animal, then circle the things that begin with the /j/ sound as in ***jug***. **Skill:** Identifying the beginning sound j.

Consonants: J

Name

Jj Jj Jj Jj Jj

My word that starts with j:

Answers will vary.

Directions: Have your child look at the signs for words that begin with *J*, then underline those words that begin with *J*. Ask your child to practice writing *Jj*, then draw and label a picture for another word that begins with *J*.
Skill: Identifying and writing Jj.

127

Beginning Sound L

Name

128

Directions: Have your child talk about the things in the picture and identify each object or animal, then circle the things that begin with the /l/ sound as in ***line***. **Skill:** Identifying the beginning sound l.

Answer Key

Directions: Have your child look at the labels for words that begin with *L*, then put an X on those labels that have *L* words. Ask your child to practice writing *Ll*, then draw and label a picture for another word that begins with *L*. **Skill:** Identifying and writing Ll.

Directions: Have your child talk about the things in the picture and identify each object or animal, then circle the things that begin with the /n/ sound as in ***news***. **Skill:** Identifying the beginning sound n.

Directions: Have your child look at the newspapers for words that begin with *N*, then underline those words that begin with *N*. Ask your child to practice writing *Nn*, then draw and label a picture for another word that begins with *N*. **Skill:** Identifying and writing Nn.

Directions: Have your child talk about the things in the picture and identify each object or animal, then circle the things that begin with the /u/ sound as in ***umbrella***. **Skill:** Identifying the beginning short u vowel sound.

Directions: Have your child look at the boxes for words that begin with *U*, then underline those words that begin with *U*. Ask your child to practice writing *Uu*, then draw and label a picture for another word that begins with *U*. **Skill:** Identifying and writing Uu.

Directions: Have your child talk about the things in the picture and identify each object or animal, then circle the things that begin with the /o/ sound as in ***octopus***. **Skill:** Identifying the beginning short o vowel sound.

ANSWER KEY

NAME

VOWELS: SHORT O

OCTOBER

Oo Oo Oo Oo

My word that starts with o:

Answers will vary.

Directions: Have your child look at the calendar for words that begin with *O*, then put an X on the days that have *O* words. Ask your child to practice writing *Oo*, then draw and label a picture for another word that begins with *O*. **Skill:** Identifying and writing Oo.

135

NAME

BEGINNING SOUND V

vases
violets

Vv

vest
vegetables

136

Directions: Have your child talk about the things in the picture and identify each object or animal, then circle the things that begin with the /v/ sound as in ***vegetables***. **Skill:** Identifying the beginning sound v.

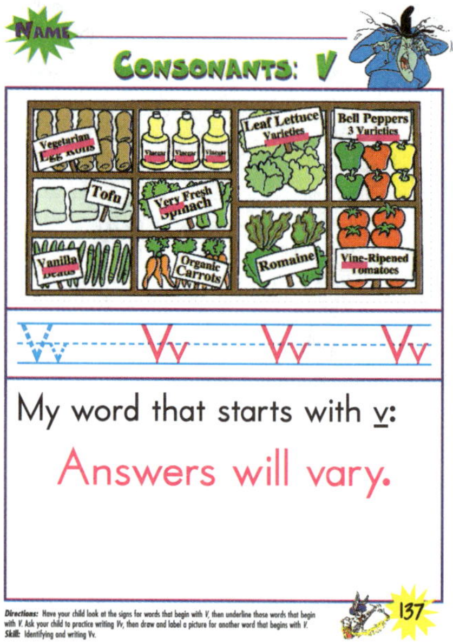

NAME

CONSONANTS: V

Vv Vv Vv Vv

My word that starts with v:

Answers will vary.

Directions: Have your child look at the signs for words that begin with *V*, then underline those words that begin with *V*. Ask your child to practice writing *Vv*, then draw and label a picture for another word that begins with *V*. **Skill:** Identifying and writing Vv.

137

NAME

PROGRESS CHECK

duck	Dd	dinosaur	lemon
iguana	Ii	hippopotamus	inchworm
kangaroo	Kk	kite	pillow
newt	Nn	mouse	nuts

138

Directions: Have your child write the uppercase and lowercase letter for the first picture in each row. Then ask your child to circle the picture in that row that begins with the same sound. **Skill:** Writing uppercase and lowercase letters; identifying beginning sounds.

NAME

PROGRESS CHECK

quail	Qq	quilt	king
wolf	Ww	violin	whale
bull	Bb	walrus	baseball bat
yak	Yy	koala	yarn

Directions: Have your child write the uppercase and lowercase letter for the first picture in each row. Then ask your child to circle the picture in that row that begins with the same sound. **Skill:** Writing uppercase and lowercase letters; identifying beginning sounds.

139

NAME

PROGRESS CHECK

fish	Ff	zipper	feather
jaguar	Jj	jug	gorilla
turtle	Tt	moose	tie
octopus	Oo	van	ox

140

Directions: Have your child write the uppercase and lowercase letter for the first picture in each row. Then ask your child to circle the picture in that row that begins with the same sound. **Skill:** Writing uppercase and lowercase letters; identifying beginning sounds.

Answer Key

NAME

PROGRESS CHECK

lion	Ll	dinosaur	ladder
zebra	Zz	inchworm	zipper
umbrella	Uu	wagon	up
fox	Xx	six	walrus

Directions: Have your child write the uppercase and lowercase letter for the first picture in each row. Then ask your child to circle the picture in that row that begins with the same sound. ***Skill:*** Writing uppercase and lowercase letters; identifying beginning sounds.

141

NAME

A FINAL WORD

alligator bull elephant goat

jaguar octopus racoon turtle

G r e a t

j o b !

142

Directions: Have your child identify the first letter in each picture word, then unscramble the letters to find the secret message. ***Skill:*** Reviewing beginning sounds and letters.

NAME

PRACTICE PAGE

Directions: Have your child use this space to practice skills learned in the previous pages.

A McGraw•Hill/Warner Bros. Workbook

The McGraw-Hill Junior Academic Series

CERTIFICATE OF ACCOMPLISHMENT

THIS CERTIFIES THAT

____

HAS SUCCESSFULLY COMPLETED
THE JUNIOR ACADEMIC'S™

Grade K Enrichment Reading

WORKBOOK.
CONGRATULATIONS AND THAT'S ALL FOLKS!

The McGraw-Hill Companies
PUBLISHER

Bugs Bunny
BUGS BUNNY, EDITOR-IN-CHIEF

McGraw-Hill Consumer Products

The skills taught in school are now available at home!
These award-winning software titles meet school guidelines and are based on
The McGraw-Hill Companies classroom software titles.

MATH GRADES 1 & 2

These math programs are a great way to teach and reinforce skills used in everyday situations. Fun, friendly characters need help with their math skills. Everyone's friend, Nubby the stubby pencil, will help kids master the math in the Numbers Quiz show. Foggy McHammer, a carpenter, needs some help building his playhouse so that all the boards will fit together! Julio Bambino's kitchen antics will surely burn his pastries if you don't help him set the clock timer correctly! We can't forget Turbo Tomato, a fruit with a passion for adventure, who needs help calculating his daredevil stunts.

Math Grades 1 & 2 use a tested, proven approach to reinforcing your child's math skills while keeping him or her intrigued with Nubby and his collection of crazy friends.

TITLE
Grade 1: Nubby's Quiz Show
Grade 2: Foggy McHammer's Treehouse

MISSION MASTERS™ MATH AND LANGUAGE ARTS

The Mission Masters™—Pauline, Rakeem, Mia, and T.J.—need your help. The Mission Masters™ are a team of young agents working for the Intelliforce Agency, a high-level cooperative whose goal is to maintain order on our rather unruly planet. From within the agency's top secret Command Control Center, the agency's central computer, M5, has detected a threat...and guess what—you're the agent assigned to the mission!

MISSION MASTERS™ MATH GRADES 3, 4, & 5

This series of exciting activities encourages young mathematicians to challenge themselves and their math skills to overcome the perils of villains and other planetary threats. Skills reinforced include: analyzing and solving real-world problems, estimation, measurements, geometry, whole numbers, fractions, graphs, and patterns.

TITLE
Grade 3: Mission Masters™ Defeat Dirty D!
Grade 4: Mission Masters™ Alien Encounter
Grade 5: Mission Masters™ Meet Mudflat Moe

MISSION MASTERS™ LANGUAGE ARTS GRADES 3, 4, & 5

This series invites children to apply their language skills to defeat unscrupulous characters and to overcome other earthly dangers. Skills reinforced include: language mechanics and usage, punctuation, spelling, vocabulary, reading comprehension, and creative writing.

TITLE
Grade 3: Mission Masters™ Freezing Frenzy
Grade 4: Mission Masters™ Network Nightmare
Grade 5: Mission Masters™ Mummy Mysteries

BASIC SKILLS BUILDER K to 2 – THE MAGIC APPLEHOUSE

At the Magic Applehouse, children discover that Abigail Appleseed runs a deliciously successful business selling apple pies, tarts, and other apple treats. Enthusiasm grows as children join in the fun of helping Abigail run her business. Along the way they'll develop computer and entrepreneurial skills to last a lifetime. They will run their own business – all while they're having bushels of fun!

TITLE
Basic Skills Builder –The Magic Applehouse

TEST PREP – SCORING HIGH

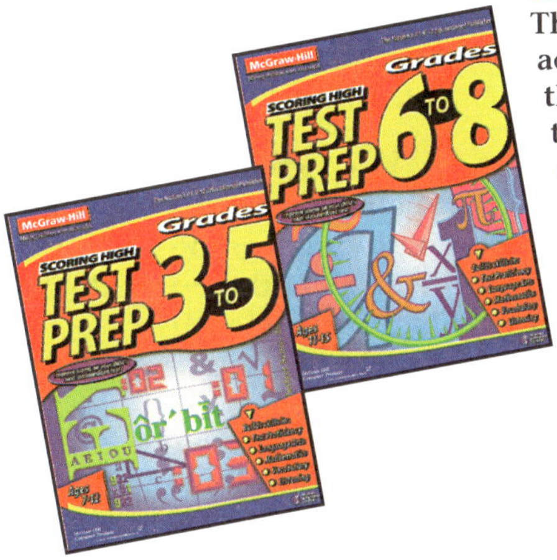

This grade-based testing software will help prepare your child for standardized achievement tests given by his or her school. Scoring High specifically targets the skills required for success on the Stanford Achievement Test (SAT) for grades three through eight. Lessons and test questions follow the same format and cover the same content areas as questions appearing on the actual SAT tests. The practice tests are modeled after the SAT test-taking experience with similar directions, number of questions per section, and bubble-sheet answer choices.

Scoring High is a child's first-class ticket to a winning score on standardized achievement tests!

TITLE
Grades 3 to 5: Scoring High Test Prep
Grades 6 to 8: Scoring High Test Prep

SCIENCE

Mastering the principles of both physical and life science has never been so FUN for kids grades six and above as it is while they are exploring McGraw-Hill's edutainment software!

TITLE
Grades 6 & up: Life Science
Grades 8 & up: Physical Science

REFERENCE

The National Museum of Women in the Arts has teamed with McGraw-Hill Consumer Products to bring you this superb collection available for your enjoyment on CD-ROM.

This special collection is a visual diary of 200 women artists from the Renaissance to the present, spanning 500 years of creativity.

You will discover the art of women who excelled in all the great art movements of history. Artists who pushed the boundaries of abstract, genre, landscape, narrative, portrait, and still-life styles; as well as artists forced to push the societal limits placed on women through the ages.

TITLE
Women in the Arts

Most titles for Windows 3.1™, Windows '95™ & '98™, and Macintosh™.

Visit us on the Internet at:

www.MHkids.com

Or call 800-298-4119 for your local retailer.

McGraw-Hill Consumer Products

All our workbooks meet school curriculum guidelines and correspond to The McGraw-Hill Companies classroom textbooks.

SPECTRUM SERIES

DOLCH Sight Word Activities

The DOLCH Sight Word Activities Workbooks use the classic Dolch list of 220 basic vocabulary words that make up from 50% to 75% of all reading matter that children ordinarily encounter. Since these words are ordinarily recognized on sight, they are called *sight words*. Volume 1 includes 110 sight words. Volume 2 covers the remainder of the list. Over 160 pages.

TITLE	ISBN	PRICE
Grades K-1 Vol. 1	1-57768-429-X	$9.95
Grades K-1 Vol. 2	1-57768-439-7	$9.95

GEOGRAPHY

Full-color, three-part lessons strengthen geography knowledge and map reading skills. Focusing on five geographic themes including location, place, human/environmental interaction, movement, and regions. Over 150 pages. Glossary of geographical terms and answer key included.

TITLE	ISBN	PRICE
Gr 3, Communities	1-57768-153-3	$7.95
Gr 4, Regions	1-57768-154-1	$7.95
Gr 5, USA	1-57768-155-X	$7.95
Gr 6, World	1-57768-156-8	$7.95

MATH

Features easy-to-follow instructions that give students a clear path to success. This series has comprehensive coverage of the basic skills, helping children to master math fundamentals. Over 150 pages. Answer key included.

TITLE	ISBN	PRICE
Grade 1	1-57768-111-8	$6.95
Grade 2	1-57768-112-6	$6.95
Grade 3	1-57768-113-4	$6.95
Grade 4	1-57768-114-2	$6.95
Grade 5	1-57768-115-0	$6.95
Grade 6	1-57768-116-9	$6.95
Grade 7	1-57768-117-7	$6.95
Grade 8	1-57768-118-5	$6.95

PHONICS

Provides everything children need to build multiple skills in language. Focusing on phonics, structural analysis, and dictionary skills, this series also offers creative ideas for using phonics and word study skills in other language arts. Over 200 pages. Answer key included.

TITLE	ISBN	PRICE
Grade K	1-57768-120-7	$6.95
Grade 1	1-57768-121-5	$6.95
Grade 2	1-57768-122-3	$6.95
Grade 3	1-57768-123-1	$6.95
Grade 4	1-57768-124-X	$6.95
Grade 5	1-57768-125-8	$6.95
Grade 6	1-57768-126-6	$6.95

SPECTRUM SERIES – continued

READING

This full-color series creates an enjoyable reading environment, even for below-average readers. Each book contains captivating content, colorful characters, and compelling illustrations, so children are eager to find out what happens next. Over 150 pages. Answer key included.

TITLE	ISBN	PRICE
Grade K	1-57768-130-4	$6.95
Grade 1	1-57768-131-2	$6.95
Grade 2	1-57768-132-0	$6.95
Grade 3	1-57768-133-9	$6.95
Grade 4	1-57768-134-7	$6.95
Grade 5	1-57768-135-5	$6.95
Grade 6	1-57768-136-3	$6.95

SPELLING

This full-color series links spelling to reading and writing and increases skills in words and meanings, consonant and vowel spellings, and proofreading practice. Over 200 pages. Speller dictionary and answer key included.

TITLE	ISBN	PRICE
Grade 1	1-57768-161-4	$7.95
Grade 2	1-57768-162-2	$7.95
Grade 3	1-57768-163-0	$7.95
Grade 4	1-57768-164-9	$7.95
Grade 5	1-57768-165-7	$7.95
Grade 6	1-57768-166-5	$7.95

WRITING

Lessons focus on creative and expository writing using clearly stated objectives and pre-writing exercises. Eight essential reading skills are applied. Activities include main idea, sequence, comparison, detail, fact and opinion, cause and effect, and making a point. Over 130 pages. Answer key included.

TITLE	ISBN	PRICE
Grade 1	1-57768-141-X	$6.95
Grade 2	1-57768-142-8	$6.95
Grade 3	1-57768-143-6	$6.95
Grade 4	1-57768-144-4	$6.95
Grade 5	1-57768-145-2	$6.95
Grade 6	1-57768-146-0	$6.95
Grade 7	1-57768-147-9	$6.95
Grade 8	1-57768-148-7	$6.95

TEST PREP
From the Nation's #1 Testing Company

Prepares children to do their best on current editions of the five major standardized tests. Activities reinforce test-taking skills through examples, tips, practice, and timed exercises. Subjects include reading, math, and language. Over 150 pages. Answer key included.

TITLE	ISBN	PRICE
Grade 1	1-57768-101-0	$8.95
Grade 2	1-57768-102-9	$8.95
Grade 3	1-57768-103-7	$8.95
Grade 4	1-57768-104-5	$8.95
Grade 5	1-57768-105-3	$8.95
Grade 6	1-57768-106-1	$8.95
Grade 7	1-57768-107-X	$8.95
Grade 8	1-57768-108-8	$8.95

Visit us on the Internet at:

www.MHkids.com

A McGraw • Hill/Warner Bros. Workbook

The McGraw-Hill Junior Academic™ Series

CERTIFICATE OF ACCOMPLISHMENT

THIS CERTIFIES THAT

____

HAS SUCCESSFULLY COMPLETED
THE JUNIOR ACADEMIC'S™

Grade K Enrichment Skills

WORKBOOK.
CONGRATULATIONS AND THAT'S ALL FOLKS!

The McGraw·Hill Companies
PUBLISHER

Bugs Bunny
BUGS BUNNY, EDITOR-IN-CHIEF

RECEIVE THE McGRAW-HILL PARENT NEWSLETTER FREE!

Thank you for expressing interest in the successful education of your child. With the purchase of this workbook, we know that you are committed to your child's development and future success. We at **McGraw-Hill Consumer Products** *would like to help you make a difference in the education of your child by offering a quarterly newsletter that provides current topics on education and activities that you and your child can work on together.*

To receive a free copy of our newsletter, please provide us with the following information:

Name ______________________________ *Store where book purchased* ______________________________

Address ______________________________ *Grade Level of book purchased* ______________________________

City __________ *State* ______ *Zip* ______ *Title of book purchased* ______________________________

e-mail (if applicable) ______________________________

Mail to:
Parent Newsletter
c/o McGraw-Hill Consumer Products
251 Jefferson Street, M.S. #12
Waldoboro, ME 04572

Or Call 800-298-4119

Or visit us at:

www.MHkids.com

This offer is limited to residents of the United States and Canada and is only in effect for as long as the newsletter is published.
The information that you provide will not be given, rented, or sold to any company.